Two Manuals Combined into One Book

Religious Support

And

Combat and Operational Stress Control

Department of Defense

Pentagon
Publishing

***FM 1-05** (FM 16-1)

Field Manual
No. 1-05

Headquarters
Department of the Army
Washington, DC, 18 April 2003

Religious Support

Contents

Distribution Restriction. Approved for public release; distribution is unlimited.

*This publication supersedes FM 16-1, 26 May 1995.

Preface

The Chaplain Corps is the Army's primary agency for practically ensuring the free exercise of religion for America's soldiers. This responsibility also applies to the Chaplain Corps' religious support to Army family members, Department of Army (DA) civilians, and other authorized personnel. The Chaplain Corps ensures free exercise of religion by executing the command religious programs of Army commanders.

PURPOSE

FM 1-05 establishes the Army's definitive doctrine for the execution of comprehensive religious support across the full spectrum of operations. It is founded upon the Army's operational doctrine articulated in FM 3-0 and FM 101-5 and other supporting documents. FM 1-05 is compatible with the joint religious support doctrine contained in JP 1-05. This doctrine pronounces that the personal delivery of religious support is the primary focus of the Chaplain Corps and places greater emphasis on the chaplain's role as an advisor to the commander. It takes into account emerging technology, possesses a broader focus on religious support at echelons above brigade, and speaks clearly to the interface between religious support operations and civil military operations. It is the authoritative, but not prescriptive, guide to the missions, roles, responsibilities, and duties the Chaplain Corps must execute to deliver comprehensive religious support.

APPLICABILITY

FM 1-05 provides guidance to commanders at all echelons responsible for providing command religious support programs to their units and organizations, as well as, to Chaplains, Chaplain Assistants and other personnel responsible to plan, prepare, and provide religious support to The Army. This guidance is the principle tool for professional education in the art of religious support operations. Trainers at all echelons should employ it as the authoritative guide for the planning and execution of religious support training. For overseas theaters, this doctrine applies to U.S. unilateral operations only, subject to applicable host nation laws and agreements.

USER INFORMATION

The proponent for this publication is the U.S. Army Training and Doctrine Command (TRADOC). Send comments and recommendations on DA Form 2028 (Recommended Change to Publications and Blank Forms) to Commandant, U.S. Army Chaplain Center and School, ATTN: ATSC-DCD, Fort Jackson, SC 29207.

Unless this publication states otherwise, masculine nouns and pronouns do not refer exclusively to men.

Chapter 1

RELIGIOUS SUPPORT FUNDAMENTALS

The First Amendment of the United States Constitution reinforced by Title 10 U.S. Code (USC) and Army Regulation (AR) 165-1, guarantees every American the right to the free exercise of religion. Commanders are responsible for those religious freedoms within their command. On behalf of the commander, Chaplains provide and perform Religious Support (RS) in the Army to ensure the free exercise of religion. Chaplains are obligated to provide for those religious services or practices that they cannot personally perform. Chaplains perform religious support when their actions are in accordance with the tenets or beliefs of their faith group. Chaplain assistants assist the chaplain in providing or performing this religious support. Religious support includes—

CONTENTS

- Religious services.
- Rites.
- Sacraments.
- Ordinances.
- Pastoral care.
- Religious education.
- Family life ministry.
- Institutional ministry.
- Professional support to the command and staff.
- Management and administration.
- Religious/humanitarian support.
- Religious support planning/operations and religious support training.

1-1. The Chaplain's statutory (10 USC) and regulatory (AR 165-1) understanding of religious support includes—

- Provide and/or perform worship services, rites, sacraments, ordinances, pastoral and spiritual care and religious education to nurture the living, care for the dying, and honor the dead.

THE CHAPLAIN'S AUTHORITY FOR MINISTRY

1-9. The government recognizes that the primary role of religious support belongs to the religious community. The federal government also recognizes the special religious support needs of armed forces personnel and the difficulty of providing organized, effective religious support. The isolation, unique environment, and training for combat zones throughout the world generate special requirements for religious support. The federal government provides the chaplaincy to meet these needs. Chaplains serve in the Army as clergy representing the respective faiths or denominations that endorse them. A chaplain's call, ministry, message, ecclesiastical authority, and responsibility come from the religious organization that the chaplain represents. Chaplains preach, teach, and conduct religious services, in accordance with the tenets and rules of their tradition, the principles of their faith, and the dictates of conscience. They also perform ecclesiastical rites and ceremonies and administer the sacraments and ordinances of their faith community. When a conflict arises between the standards of a chaplain's faith tradition and the requirements of a religious support mission, chaplains are required to provide for the religious needs of the soldier by obtaining other chaplains or qualified people to perform the needed religious support.

1-10. Chaplains, ordained and endorsed by distinctive faith groups, must be people of integrity who consistently live according to the highest principles of the Army and the distinctive faith groups they represent. Chaplain assistants are also expected to be people of integrity whose lives reflect high moral standards and the values of the Army. The UMT must mirror the Army's values – Loyalty, Duty, Respect, Selfless Service, Honor, Integrity, and Personal Courage.

1-11. As religious leaders, chaplains continue to meet the vision of General George Washington by setting the standard for moral and ethical behavior by upholding the Army Chaplaincy Values: Spirituality, Accountability, Compassion, Religious Leadership, Excellence, and Diversity (SACRED).

Army Chaplaincy Values
Spirituality
The bedrock of all faiths, providing meaning and direction in life.
Accountability
The means to measure how we practice what we preach.
Compassion
The love that sees beyond the surface and discerns our common humanity.
Religious leadership
Interprets the role of leadership prophetically.
Excellence
The standard for staff work and religious responsibilities.
Diversity
Being respectful of different views and ideas that are not like our own.

RELIGIOUS SUPPORT MISSION

1-12. The mission of the UMT is to provide and perform religious support to soldiers, families, and authorized civilians as directed by the commander. Chaplains serve as personal staff officers to commanders at all levels of the command providing essential information on troop and unit morale, quality of life matters, free exercise of religion issues, ethical decision-making, and the impact of religion on the operation.

RELIGIOUS SUPPORT FUNCTIONS

1-13. Comprehensive religious support includes the following three major functions:

- ***Nurture the Living.*** In preparation for missions that span the full spectrum of operations UMTs develop and provide religious support activities to strengthen and sustain the spiritual resilience of soldiers and family members. During the battle, UMTs bring hope and strength to those who have been wounded and traumatized in body, mind, and spirit, assisting the healing process.
- ***Care for the Dying.*** UMTs provide religious support, spiritual care, comfort, and hope to the dying. This focus of religious support affirms the sanctity of life, which is at the heart of the chaplaincy. Through prayer and presence, the UMT provides the soldier with courage and comfort in the face of death.
- ***Honor the Dead.*** Our Nation reveres those who have died in military service. Religious support honors the dead. Funerals, memorial services, and ceremonies, reflect the emphasis the American people place on the worth and value of the individual. Chaplains conduct these services and ceremonies, fulfilling a vital role in rendering tribute to America's sons and daughters who paid the ultimate price serving the Nation in the defense of freedom.

RELIGIOUS SUPPORT ACTIVITIES

1-14. Commanders provide opportunities for the free exercise of religion. UMTs provide and/or perform the following major religious activities, within the unique military context, in support of the command religious plan:

- ***Religious Services:*** Worship; Funerals; Memorial Services and Ceremonies; Prayer Breakfasts.
- ***Rites, Sacraments, Ordinances:*** Marriages; Burials; Baptisms; Confirmations; Blessings; Daily Prayers; other ministrations and religious requirements.
- ***Pastoral Care/Counseling:*** Taking part in command activities; visiting soldiers; calling on families; activities and unit "ministry of presence"; individual and group pastoral counseling; and similar pastoral activities. This may also include elements of the spiritual fitness training program. .
- ***Religious Education:*** Study of religious teachings including activities of faith sustainment (enrichment, formation, study or instruction), marriage enrichment, problem solving, communication skills, parenting skills, youth programs, religious literature.

- ***Family Life Ministry:*** Training UMTs, directing the Chaplain Family Life Center (CFLC). Family Life Ministry is implemented through the CFLC programs and formal training.
- ***Institutional Ministry:*** Preserving soldiers' fighting strength by maintaining their health and supporting family members receiving medical services. Institutional ministry occurs in hospitals and confinement or correction facilities.
- ***Professional Support to the Commander and Staff:*** Support to the commander on matters of religion, morals, and morale, as affected by religion and the impact of indigenous religions on the military mission. Assesses the unit climate by visiting soldiers and providing advice on the religious needs of soldiers and families.
- ***Management and Administration:*** Beyond the normal administration requirements and actions required by the Command Religious Plan, the UMT manages funds, facilities, manpower, supplies, and equipment.
- ***Religious/Humanitarian Support:*** Support programs to Joint Military Operations. Such support includes humanitarian support programs on issues of religion, morale, morals, and ethics.
- ***Religious Support Planning/Operations:*** Providing commanders with Religious Support plans and programs for all mobilization requirements spanning the full spectrum of military operations, as well as, planned ministries for families of deployed soldiers.
- ***Religious Support Training:*** Training that includes spiritual fitness (Spiritual fitness training is the training and development of personal qualities needed to sustain a soldier in times of danger, hardship, and tragedy. These qualities are based in religious, philosophical and human values and they form the basis of character, decision-making and integrity.), moral leadership (see AR 165-1, Chapter 11), soldier and family member suicide prevention, violence in the work place, professional leadership, and training required by UMTs, commanders, staff, soldiers and family members.

ROLES AND RESPONSIBILITIES

1-15. Commanders, chaplains, and chaplain assistants have distinct roles and responsibilities in providing and performing religious support to soldiers, their family members and authorized civilians.

COMMANDER

1-16. The commander provides for the free exercise of religion of soldiers and families thereby enhancing the spiritual health of the command. To assure readiness across the full spectrum of operations, commanders provide for a wide range of religious support activities. Commanders use the Commanders' Religious Program (CRP) to resource the religious support plan.

1-17. The commander accommodates religious practices, in accordance with Army regulations (See AR 600-20). The Army policy is to accommodate religious practices when those accommodations will not have an adverse impact on military readiness, unit cohesion, standards, health, safety or discipline, or otherwise interfere with the performance of the soldier's military duties. Accommodation of a soldier's religious practices must be weighed against military necessity and cannot be guaranteed at all times.

CHAPLAIN

1-18. Chaplains personally deliver religious support. They have two roles: religious leader and staff officer.

- The chaplain as a religious leader executes the religious support mission, which ensures the free exercise of religion for soldiers and authorized personnel.
- The chaplain is a non-combatant and will not bear arms (AR 165-1).

> The Chaplain is a non-combatant and will not bear arms in accordance with AR 165-1.

1-19. The chaplain is a personal staff officer responsible for coordinating the religious assets and operations within the command. The chaplain is a confidential advisor to the commander for religious matters. A chaplain is located at every echelon of command above battalion. Besides his common staff responsibilities, the chaplain's specific responsibilities include —

- Advise the commander on issues of religion, ethics, and morale (as affected by religion), including the religious needs of all personnel for whom the commander is responsible.
- Provide commanders with pastoral care, personal counseling, advice, confidentiality, and sacred confidence.
- Develop and implement the commander's religious support program.
- Exercise staff supervision and technical control over religious support throughout the command.
- Provide moral and spiritual leadership to the command and community.
- Coordinate religious support with unit ministry teams of higher and adjacent headquarters, other services, and multinational forces or coalition partners.
- Translate operational plans into battlespace ministry priorities for religious support.
- Help the commander ensure that all soldiers have the opportunity to exercise their religion.
- Advise the commander and staff, with the G5 (S5), on the impact of the faith and practices of indigenous religious groups in an area of operation.
- Perform or provide religious rites, sacraments, ordinances, services, and pastoral care and counseling to nurture the living, care for the dying and honor the dead.
- Provide religious support to the command and community to include confined or hospitalized personnel, enemy prisoners of war (EPWs), civilian detainees, and refugees.
- Support the commander in the execution of Civil Military Operations (See appendix A).

1-20. The Chaplain trains, equips and supports the subordinate chaplain and chaplain assistant (FM 101-5).

1-21. As a minimum, the chaplain requires a SECRET clearance. This will allow the chaplain to have access to the command post and enable him/her to be part of the planning process.

1-22. Although chaplains do not have command or Uniform Code of Military Justice (UCMJ) authority, they do exercise staff supervision and functional direction of religious support personnel and activities.

1-23. The proper title for a chaplain is "chaplain" regardless of military rank and professional title.

CHAPLAIN ASSISTANT

1-24. The chaplain assistant (CA) is a combatant, trained in religious support and staff skills; who possesses a SECRET clearance; manages force protection, provides security for the UMT on the battlefield, and accompanies the chaplain in the area of operation. The CA is a soldier trained to assist the chaplain in providing religious support. Under the direction of the chaplain, the CA coordinates and synchronizes religious support operations. In the absence of the chaplain, the CA continues the religious support mission for the commander as a staff representative, but does not assume the religious leadership role of the Chaplain.

1-25. Unit Ministry Team Chaplain Assistant. Under the supervision of the chaplain, the CA—

- Assists in religious support planning and operations.
- Coordinates the UMT movement plan, monitors the tactical situation, and attends briefings.
- Coordinates and prepares for religious activities.
- Coordinates for and conducts equipment maintenance.
- Assesses soldier morale and provides this information to the chaplain.
- Conducts pre-counseling interviews.
- Assists with combat operations stress response (COSR), identification, intervention, and care.
- Performs tactical, logistical, and administrative support for the UMT.
- Requires access to classified messages, systems, and secure areas to accomplish the religious support mission.
- When an NCO, the CA serves as a liaison for the chaplain to the noncommissioned officers (NCOs) of the command.

> Chaplain Assistants require a SECRET clearance to accomplish the religious support mission.

1-26. The Chaplain Assistant NCO supervises, mentors, and trains chaplain assistants and serves as the religious support section noncommissioned officer in charge (NCOIC) according to the echelon assigned. In addition to executing the roles and functions of a CA of lesser rank, the NCOIC performs many of the following staff functions:

- Participates in battle staff planning, tracking, and execution.
- Procures contracts for goods and services, and coordinates and monitors the CRP.

- Provides training in suicide prevention awareness, intervention for COSR, and performance counseling.
- Plans, briefs, and performs operational training with other staff NCOs.
- Coordinates taskings.
- Serves as advisor to the unit CSM and other key NCO leaders.
- Coordinates religious support during joint, multinational, and interagency operations.

ORGANIZATION

UNIT MINISTRY TEAM AND MODIFIED TABLE OF ORGANIZATION AND EQUIPMENT (MTOE) UNITS

1-27. UMTs authorized by the Modified Table of Organization and Equipment (MTOE) are assigned to units whose primary mission is warfighting. The UMT deploys with its unit and provides religious support for all units in the commander's area of responsibility during each stage of force projection.

1-28. At echelons above brigade, the senior chaplain is at times listed in the command section of the MTOE. Additional chaplains, chaplain assistants, and UMT equipment are listed in a separate section of the MTOE.

1-29. Comprehensive religious support requires the assignment and deployment of UMTs in maneuver, maneuver support, and maneuver sustainment units at all echelons, beginning at the battalion. UMTs move forward to provide religious support to all elements of the squadron/battalion: companies, troops, batteries, platoons, squads, and teams.

1-30. In operations where lethality and intensity are high, religious support becomes increasingly important to soldiers. UMTs provide hope, faith, encouragement, and compassion to soldiers experiencing shock, fatigue, isolation, fear, and death. In the chaos and uncertainty of military operations, the UMT is a reminder of the presence of God.

UNIT MINISTRY TEAMS AND TABLE OF DISTRIBUTION AND ALLOWANCES (TDA) UNITS

1-31. UMTs authorized by a TDA are assigned to units whose primary missions vary widely. The missions and responsibilities associated with TDA organizations depend upon the type of organization served. TDA organizations include installations, schools, hospitals, disciplinary barracks, some Major Army Commands (MACOMs), and Headquarters, Department of the Army (HQDA). The Installation UMT provides the training base for all MTOE and TDA UMTs. In addition to providing religious support directly to the units, the TDA UMTs manage, supervise, and support the religious support mission of their organization.

PROVIDING UNIT RELIGIOUS SUPPORT

1-32. Unit religious support is provided in the following three ways:

- ***Unit support:*** support provided to the unit to which the UMT is assigned or attached. The team normally gives first priority to this mission.

- ***Area support:*** support provided to soldiers, members of other services, and authorized civilians who are not a part of the team's unit, but operate within the supporting unit's area of operations (AO).
- ***Denominational support:*** support given to soldiers and other authorized persons of the chaplain's denomination or distinctive faith group. Limited assets affect the availability of denominational support.

1-33. The UMT extends opportunities for soldiers to serve as volunteers in worship, religious education, and service to the unit. The sponsoring UMT is responsible for recruiting, training, and supervising volunteers. Under the supervision of the unit chaplain, soldiers may serve as members of their own faith groups as Distinctive Faith Group Leaders (DFGL). Denominational certification, the approval of the commander, and the final approval of the MACOM/State Area Command (STARC)/United States Army Reserve Command (USARC) or Major Subordinate Command (MSC) chaplain are required before a soldier may serve as a DFGL. (See AR 165-1)

SUMMARY

1-34. The Chaplain Corps and Regiment remains committed to providing for the fundamental freedom of the free exercise of religion as guaranteed by the Constitution. This is accomplished by—

- Performing religious activities according to the chaplain's faith tradition.
- Providing professional and effective advice to the commander, in coordination with commander's staff, regarding the needs of soldiers not of the chaplain's faith tradition.
- Developing effective teamwork within the UMT.
- Through cooperation with other chaplains without compromising the tenets of one's own faith.

1-35. Through it all, the soldier's spiritual health and welfare, as well as the soldier's spiritual freedoms, remain the focal point. Thus the Chaplain Corps' unofficial motto: *"Bringing God to Soldiers, and Soldiers to God."*

RESOURCES

1-36. The following resources were used for writing and updating this doctrine.

- AR 165-1.
- AR 600-20.
- JP 1-05.
- JP 5-00.2.
- JP 3-0.
- FM 1 (100-1).
- FM 3-0 (100-5).
- FM 16-1.
- FM 22-100.
- DA Pam 165-3.
- Military Rules of Evidence 503 UCMJ.
- U.S. Army Chaplaincy Strategic Plan FY 2000-FY 2005.
- Chaplain Support Force XXI White Paper.
- Chaplain Assistant Concept Paper, 2000.

Chapter 2

RELIGIOUS SUPPORT ENVIRONMENT

"We have to remember something simple but important: Making history is messy. Studying Normandy today looks like arrows and unit symbols. Normandy on 6 June 1944 looked like Saving Private Ryan – dangerous and chaotic."

MG James Dubik

The mission and operational environment determine the shape of religious support. The team conducts religious support effectively with a clear understanding of the context and complexities of military operations in which religious support is provided and performed.

CONTENTS

RELIGIOUS SUPPORT AND THE THREAT ENVIRONMENT

2-1. With the collapse of the Soviet Union, the bilateral world of the post WWII era, pitting the political ideologies of communism and capitalism against one another, came to an end. As a consequence, the operational climate and the organizational structure of the U.S. Army began to change. The Army moved away from the concept of a forward deployed Army based overseas to the concept of a force projection Army based in the United States. "Because the Army is a power projection force, its units deploy from and return to home stations." (FM 3-0, pg. 4-9) The changes in operational climate, organization, and deployment posture require adaptations in the methods of providing religious support.

THE THREAT

"The enemies we are likely to face through the rest of the decade and beyond will not be 'soldiers,' with the disciplined modernity the term conveys, but 'warriors' – erratic primitives of shifting allegiance, habituated to violence, with no stake in civil order."

LTC Ralph Peters, Fighting for the Future

2-2. Potential threats to U.S. forces are many and varied. They range from a local populace with small arms to a conventional force possessing weapons of mass destruction (WMD). Global, regional, and local challenges confront the U.S. Army with adversaries who, as a result of their religious beliefs, may interpret such fundamental ideas as right and wrong, the value of human life, or even the concept of victory and defeat very differently.

2-3. Failing states and the associated fragmentation due to religious or ethnic strife will continue to promote regional instability and produce security concerns for the United States and our allies. The security environment is further complicated by challenges that transcend state borders, such as ethnic disputes and religious rivalries.

2-4. Threats can be of short duration or protracted over a long period of time. They can be waged over competing ideology, ethnic identity, economic policies, national aspirations, religious conflict or any number of social and political issues. Struggles can be waged by a variety of means: diplomatic, informational, military, or economic.

2-5. Asymmetric threats (unconventional or inexpensive approaches) to the United States (such as terrorism—both domestic and international, guerilla warfare, and WMD including attacks on domestic targets) are issues of special concern and create challenges to religious support planning. Particularly problematic is the growing threat to our increasing information infrastructure through acts of cyber-terrorism.

2-6. Unlike the Cold War era, when threats were measurable and predictable, Army forces today are likely to encounter conditions of greater ambiguity and uncertainty. The U.S. Army must be prepared to respond to worldwide strategic, operational and tactical challenges either alone, or as part of a joint or multinational team. The U.S. Army's UMTs must also be prepared to provide religious support in every contingency or circumstance.

2-7. Mission, enemy, terrain and weather, troops and support available, time available, civilian considerations (METT-TC) includes "civilians" as an operational planning consideration. The UMT requires skills to deal with the religious support requirements of DA civilians, contractors in the area of operations, and indigenous and host nation civilians. With the expected increased urban terrain missions, the UMT will have more opportunity and responsibility to interface with the local people and religious leaders as it supports the command mission.

FORCE PROJECTION

Force Projection Process	
Mobilization	Sustainment
Deployment	Redeployment
Employment	Demobilization

2-8. The Army performs its mission within the context of national security and military strategies. Power projection is the application of the elements of national power—diplomatic, economic, information, and military—anywhere in the world.

2-9. The military aspect of power projection is force projection. Force projection is the deployment of forces to perform missions spanning the whole range of military operations anywhere in the world. Force projection applies to the entire Army, Active and Reserve Components.

2-10. Force projection encompasses a process in which forces are managed, trained, and projected to the theater of operations for any contingency. The force projection process is organized in phases: Mobilization, Deployment, Employment, Sustainment, Redeployment, and Demobilization.

2-11. Credible force projection depends on the capability of the United States to deploy forces rapidly and to perform missions spanning the whole range of military operations. A unit may deploy with short notice, causing tremendous stress on soldiers and systems.

2-12. When the United States decides to employ its military resources, the nation continues to meet the soldier's rights to free exercise of religion by deploying UMTs with the soldiers. During force projection operations, the religious support mission is to provide religious support to the soldiers, authorized civilians, and families of the deployed unit. The environment and type of operation may change, but the religious support mission remains constant throughout the spectrum of operations. The successful provision of religious support to meet requirements of force projection rests upon four pillars (table 2-1).

Table 2-1. Four Pillars of Religious Support

Religious support plan (RSP) {to include the CMRP}	Coordinated with all echelons of command and included in the operation plan/order.
Communication architecture	UMTs must be able to interact with their commands and RS providers at higher, subordinate and adjacent units. (Plan redundancy when possible).
Split-based operations	Concurrent RS provided from projection platform to theater of operations.
Force tailoring	UMTs assigned to operational units and Task Forces to meet unique religious support requirements, liaison with faith-based Non-Governmental Organizations (NGOs) and International Organizations (IOs), and governmental agencies.

LEVELS OF MILITARY ACTIONS

2-13. Levels of military actions include strategic, operational, tactical, and military operations other than war (MOOTW). Military operations in the 21st century will span all levels of military actions, often simultaneously or in rapid sequence. For example, forces on the ground may be engaged in high-intensity conflict one moment, then with changes in the strategic diplomatic or political circumstances, be leading a MOOTW operation the next moment, or vice versa. The Army Chaplaincy of the 21st Century will be organized and equipped to meet the challenges at all levels of actions at any time.

STRATEGIC ENVIRONMENT

"Although U.S. military forces enjoy superiority in many dimensions of armed conflict, the United States is likely to be challenged by adversaries who possess a wide range of capabilities, including asymmetric approaches to warfare, particularly weapons of mass

destruction. The United States cannot predict with a high degree of confidence the identity of the countries and actors that may threaten its interests and security."

--Quadrennial Defense Review Report, 2001

Strategic Religious Support Mission Essential Task List

2-14. The Army's strategy is to achieve and maintain a capabilities-based, threats-adaptive Army that is postured to support our nation's strategy to respond to the full spectrum of crises. To support the Army's strategy, the Chaplaincy must anticipate changes in the strategic environment, technological advances, and evolving Joint war-fighting concepts. The Army's strategic mission essential task list (METL) describes what "well–trained, superbly led, and well-equipped soldiers do for the nation (FM 3-0)." Likewise, the Army Chaplaincy in support provides strategic METL tasks that shape and sustain the spiritual freedoms of the soldier.

Army and Chaplain Corps Mission Essential Task Lists

Army—

- Shape the security environment.
- Respond promptly to crisis.
- Mobilize the Army.
- Conduct forcible entry operations.
- Dominate land operations.
- Provide support to civil authorities.

Chaplain Corps—

- Provide and perform personal delivery of religious support throughout full spectrum of operations.
- Train and equip the Chaplain Corps.
- Advise the commander on matters pertaining to the religious, moral, ethical, and morale climate of the command.
- Advise the commander on matters pertaining to indigenous religions and their impact on military operations.
- Manage mobilization of the Chaplain Corps throughout all Army components.
- Plan and coordinate for religious support with other services, international chaplaincies, and governmental agencies.
- Provide UMTs to implement Combatant Commanders' religious support requirements. (Combatant Commanders are the commanders of the nine unified commands in DOD i.e. USEUCOM, USPACOM, USCENTCOM, USSOUTHCOM, USSOCOM, USTRANSCOM, USJFCOM, USNORTHCOM, USSTRATCOM)
- Provide and perform religious support during military support to civilian authorities (MSCA).

OPERATIONAL ENVIRONMENT

2-15. The operational level of war links the tactical employment of forces to strategic objectives. The focus at this level is on operational art—the use of military forces to achieve strategic goals. Emerging doctrine from all branches is shaping the religious support environment by making increasing requirements for support in this much wider spectrum of operations.

2-16. On the nonlinear, asymmetrical battlefield, with operations characterized by multiple and simultaneous operational missions, religious support tasks become much more challenging. A sound operational religious support plan, executed skillfully, increases the chances of tactical success. It does this by decreasing the tactical UMTs risks and supports the UMT's time management requirements. Operational religious support planning integrates conditions, sequences of support actions, and wise management of resources throughout the theater, to optimize the delivery of religious support.

TACTICAL ENVIRONMENT

2-17. Tactics is the art and science of employing all available means to win battles and engagements. Tactical operations are rapid and dynamic in nature. Specifically, the subject of tactics comprises the actions taken by a commander to arrange units and activities in relation to each other and the enemy. At this level, UMTs are fully engaged with their units maneuvering to the area of operations delivering religious support.

MILITARY OPERATIONS OTHER THAN WAR ENVIRONMENT

2-18. Military Operations Other Than War (MOOTW) includes a wide range of contingencies. It may involve offensive or defensive operations, but most often the primary focus is on stability operations and support operations. During MOOTW, UMTs focus their religious support planning and execution on spiritual care of the soldier over long durations of deployment and serve, when directed, as the commander's liaison with local religious leaders, and faith-based NGOs and IOs.

DECISIVE FULL SPECTRUM OPERATIONS

2-19. Within the operational and tactical levels of military actions, commanders plan and execute military operations. There are four types of operations: Offense, Defensive, Stability and Support. The Army's role differs with each operation (Figure 2-1).

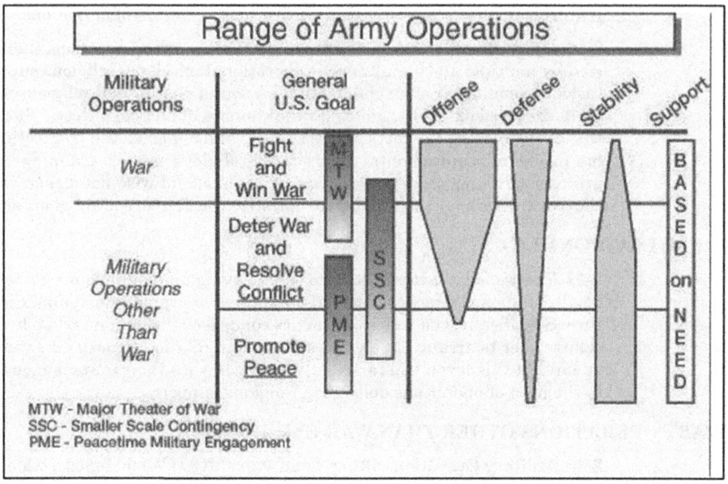

Figure 2-1. Range of Army Operations

OFFENSE

2-20. This is the decisive form of war. Offense is essential to success in all operations—defensive, stability, and support as well as offensive.

DEFENSIVE

2-21. These operations are directed by command to defeat enemy attacks, buy time, economize forces, or develop conditions favorable for the offense.

STABILITY

2-22. Stability operations include a range of actions that Army forces conduct outside the United States and its territories. The main goal is to promote and sustain regional and global stability. Characteristics of stability operations include a diverse and complex environment. These operations are highly demanding on small units and leaders, require intense mental and physical agility, and are often long-term. There are ten types of stability operations:

- Peace operations.
- Foreign Internal Defense.
- Security assistance.
- Humanitarian and civic assistance.
- Support to insurgencies.
- Support to counter drug operations.
- Combating terrorism.

- Noncombatant evacuation operations.
- Arms control.
- Show of force.

SUPPORT

2-23. Support operations are conducted to relieve suffering and help civil authorities prepare for, or respond to, crises. They are divided into two categories: Domestic support operations (DSO) conducted within the United States and its territories, and foreign humanitarian assistance (FHA) conducted outside the United States and its territories. Support Operations usually are conducted to relieve or reduce the results of natural or manmade disasters. They are also conducted to relieve conditions (such as pain, disease, hunger, or privation) that present a serious threat to life or loss of property. There are several forms of support operations.

- ***Relief operations.*** Humanitarian relief focuses on the well being of supported populations. Disaster relief focuses on recovery of critical infrastructure after a natural or manmade disaster.
- ***Support to incidents involving weapons of mass destruction (WMD).*** Assistance is given to civil authorities protecting U.S. territory, population, and infrastructure before an attack by supporting domestic preparedness and protection of critical assets.
- ***Support to Civil Law Enforcement.*** This involves activities related to counter terrorism, counter drug operations, military assistance to civil disturbances, and general support.
- ***Community assistance.*** Provision is given to community assistance for projecting a positive military image, training opportunities, and public relationships.

Religious Support in Peace Operations

2-24. The range of religious support is tailored to support the objectives outlined in the operation plan or order. Unit Ministry Teams must understand each of the four types of military operations in order to plan for, and execute, religious support.

2-25. The Army performs peace operations that range from explicitly peaceful missions to those that have significant applications of combat power. Peace operations are classified as peace making, peace building, and preventative diplomacy. These operations are conducted outside U.S. territories (FM 100-23).

Religious Support in Joint, Combined, and Interagency Operations

2-26. Army units deploy with other military services and agencies of the U.S. government. Increasingly, the U.S. Army operates with other countries as well. Army forces will normally deploy as part of a joint, multinational, and interagency operation. These operations are collectively called unified actions. This is true not only when the military effort is the prime strategic option, but also when other strategic options are preferred.

2-27. Joint operations are operations in which two or more American military departments are employed in the same operation. Joint doctrine is published under the authority of the Chairman of the Joint Chiefs of Staff.

JP 1-05 provides doctrinal guidance for religious support in joint operations. It is important to recognize that Joint publications take precedence over doctrine of the participating service branches.

2-28. Combined operations are operations in which American forces are employed along with forces from other nations. Operational procedures are determined through agreement and coordination by higher headquarters. Provisions for religious support should be included in the operational procedures or policies agreed upon by the participating nations and coalitions; for example, American, British, Canadian, and Australian/New Zealander (ABCA) or the North Atlantic Treaty Organization (NATO).

2-29. Interagency operations are operations in which the military works in tandem with another government agency. The lead agency and organizational relationship are specified in the mission directive. The provision of religious support must be specified for each operation.

2-30. Standardized methods of operating may not be present in many Joint, Combined, or Interagency Operations. Circumstances will occur that cannot be anticipated. UMTs must be flexible and creative in planning religious support for these operations. In each type of operation, the key elements to consider for religious support mission accomplishment are understanding, cooperation, and coordination. These will take place between U.S. chaplains, multinational chaplains, and interagency representatives. This is essential to achieve unity of effort in providing total and comprehensive theater of operation religious support.

Religious Support in Special Operations

2-31. Special Operations Forces (SOF), which consist of specialized Army, Air Force, and Navy units, give the President and the Secretary of Defense additional military options across the full spectrum of operations. These options provide a strategic response capability to prevent future, or stabilize current, international situations with minimum risk to U.S. interests. The theater SOF commander executes special operations (SO) as part of the theater commander's joint special operations effort. Religious support is woven into this specialized Army and joint environment. The religious support planning encompasses all contingencies, whether direct or indirect military actions in support of a Joint Force Command.

2-32. U.S. Army Special Operations Command (USASOC) commands all CONUS-based active and reserve Army Special Operations Forces (ARSOF) units. This command is comprised of five types of units: Special Forces (SF), Rangers, Army Special Operations Aviation (ARSOA), Psychological Operations (PSYOP), and Civil Affairs (CA).

2-33. The command is responsible for readiness and training oversight for all Army Reserve SOF and all outside the continental United States (OCONUS) forward based (deployed) active Army SOF. When directed, USASOC provides mission-ready ARSOF assets, through United States Special Operations Command (USSOCOM), to the regional combatant commanders for employment under their command authority. (See Chapter 5 for religious support tasks.)

Religious Support and Military Support to Civilian Authorities

2-34. Another military environment in which religious support is provided is Military Support to Civilian Authorities (also known as MSCA). These domestic support operations involve the Army in assisting civilian authorities by providing general military assistance. General assistance for domestic support operations includes disaster assistance, environmental assistance, and law enforcement support and community assistance. These operations are conducted within U.S. territories (FM 100-19).

2-35. Religious support in both peace operations and MSCA, as well as Military Assistance to Civil Disturbances (MACDIS), require intentional coordination and cooperation within the Chaplaincy among Army components (active duty, Reserve, and National Guard (NG)), as well as with civil and NGO leaders.

2-36. Civil Support will involve extensive coordination and liaison among interagency, joint, multi-jurisdictional (state and local), and active and Reserve Component (AC/RC) chaplaincies. When state authorities request Federal help, the Federal Emergency Management Agency (FEMA) most often is the lead agency, with DOD supporting. The Secretary of the Army is the executive agent. The Army Chaplaincy works directly for the military command providing RS to soldiers and authorized civilians.

2-37. Reserve Component UMTs provide a unique capability for the Chaplaincy. They have a long-term relationship with state and local religious leaders, institutions, officials, and businesses. They also have special capabilities and are geographically dispersed to conduct religious support operations. UMTs provide and perform the following religious support before, during, and after crises:

- *Planning*. Army National Guard (ARNG) and Reserve component (RC) UMTs are ideally suited to support any locality because of their dispersion throughout the Nation and long-term relationships with local clergy, civil agencies, and institutions. Planning must consist of information management, synchronization with all military and civil assets, equipping and training for all contingency missions, close support of Civil Military Operations tasks, shaping the force religious support structure to meet operational needs and staff integration.
- *Preparing*. Doctrine, organization, training, material, leader development, personnel and facilities (DOTMLPF) issues need to be continuously reviewed by chaplain leadership in all Army components to respond effectively within Civil Support operations. All written guidance and plans need to be continuously updated to meet METT-TC needs and directives. Standard operating procedures (SOPs), UMT battle-books, and other publications need to address all areas to which all UMTs will be directed to respond. Training incorporates many of the collective skills that UMTs already possess and continually update.
- *Responding and executing*. When a crisis occurs, UMTs of selected units, in support of the lead agency, will provide and perform direct religious support to military units, will perform liaison responsibilities with State and Federal agencies pertaining to religious issues, will conduct religious risk assessment for the command, and will liaison with civilian religious leaders.

- *Follow-up*. In the days following a crisis, UMTs provide and perform religious support directly to military forces as they are employed after a hostile event, to include the following tasks: mass casualty religious support, critical event debriefings, liaison with local religious leaders, etc.

2-38. The Medical Command (MEDCOM) plays a key role in response efforts. Existing MEDCOM units may be tailored to meet the METT-TC requirements. MEDCOM UMTs will bring specific skills in mass casualty, trauma, crisis intervention, and critical event debriefing ministry to these events. RS planning must be well integrated within this health care response. Army AC/RC deployable hospitals may move forward to the area of operations. Military medical facilities may be used as a trauma center when that capability does not exist locally, or when called for under National Disaster Medical System (NDMS). Large numbers of casualties resulting from WMD will challenge the UMT and other caregivers, resulting in increased need for critical event debriefing and other types of intervention ministries. Special Medical Augmentation Response Teams–Pastoral Care (SMART-PC) play a unique religious support role in response ministry. SMART-PCs, as well as the chaplain's role in support of Special Medical Augmentation Response Teams–Stress Management (SMART-SM), are discussed in depth in chapter 6 of this FM.

SUMMARY

2-39. Religious support provided during a military operation can occur under widely differing circumstances. The UMT's basic mission to provide religious support to the soldier, however, remains constant. What differs is the operational environment in which the religious support is provided. The differing types of military operations determine the specific nature of religious support.

2-40. The U.S. Army's UMTs must be prepared to provide religious support in every contingency or circumstance. With the expected increase in urban terrain missions, the UMT will have more opportunity and responsibility to interact with the local people and religious leaders in supporting the command mission. Additionally, METT-TC includes "civilians", which include both authorized DA Civilians, civilian contractors on the battlefield, as well as indigenous and host nation civilians. These civilians are included as an operational planning consideration for the UMTs to determine who, what, when, and where to provide and perform religious support.

2-41. In all types of operations across the full spectrum, key elements to consider for the religious support mission accomplishment are: understanding, cooperation, and coordination between U.S. chaplains, multinational chaplains, and interagency representatives. This is essential to achieve unity of effort in providing total and comprehensive theater of operation religious support.

2-42. There are no boundaries between strategic, operational, tactical, and MOOTW environment. Over time in the battlespace of a theater, operational activities link the tactical employment of forces to strategic objectives. The Army strategic METL describes what "well –trained, superbly led, and well-equipped soldiers do for the nation (FM 3-0)." Likewise, the Army

Chaplain Corps, in support, provides a strategic METL that helps shape and sustain the spiritual center of the soldier.

2-43. Civil Support will involve extensive coordination and liaison among interagency, joint, multi-jurisdictional (state and local), and active and reserve component (AC/RC) Chaplaincies. The RC UMT provides a unique capability to the Chaplaincy. They have long-term relationships with state and local religious leaders, institutions, officials, and businesses. Also, RC UMTs have special capabilities and geographic dispersion to conduct religious support operations. Religious Support tasks will be in concert with the Army's role in Civil Support which follow in these categories: force protection, support to crisis management, support to consequence management, protection of critical assets, support to counter-terrorism, deterrence/defense against strategic attack, and MSCA missions.

RESOURCES

2-44. The following is a list of main resources utilized for writing and updating this doctrine.

- AR 165-1.
- JP 1-05.
- JP 5-00.2.
- JP 3-0.
- JP 3-05.
- FM 3-0 (100-5).
- FM 16-1.
- FM 22-100.
- FM 1 (100-1).
- FM 100-19.
- National Military Strategy
- Joint Vision 2010
- Army Vision 2010
- A National Security Strategy For a New Century, May 1997
- U.S. Army Chaplaincy Strategic Plan FY 2000-FY 2005
- Chief of Chaplains MSCA Concept Paper, September 2000
- Chaplain Support Force XXI White Paper
- Quadrennial Defense Review Report, September 2001.

Chapter 3

UNIT MINISTRY TEAM OPERATIONS

"I was determined that whatever else might happen next time [post-W.W.I], at least I myself would be prepared, and trained, and ready when the time came. I had at times a kind of inward feeling that the call would come, to me personally, and in my prayers morning and evening I used to ask that I might be given help and strength so that I might not fail when put to the test."

--Field Marshall Montgomery, The Path of Leadership

"Everything is simple in war, but the simplest thing is difficult."

--Carl Von Clausewitz, On War

This chapter establishes the essential elements of Religious Support (RS) planning, coordination, and execution efforts in all phases of operations. It examines the following key points:

- General Religious Support planning, preparation, coordination and execution overview.
- Religious Support Planning for Force Projection.
- General Responsibilities of the UMT During Religious Support Operations.

CONTENTS

GENERAL RELIGIOUS SUPPORT PLANNING, PREPARATION, COORDINATION AND EXECUTION OVERVIEW

3-1. To meet the religious needs of soldiers and other authorized personnel in all operations, the Unit Ministry Team (UMT) prepares a comprehensive religious support plan that is staffed and integrated into every operational phase. The UMT advises the commander on the plan for accomplishing the religious support mission. The plan encompasses RS required from training and mobilization, to redeployment and demobilization.

3-2. The commander is responsible for the execution of the Religious Support Plan (RSP). The Chaplain, as a member of the commander's personal staff, is responsible to the commander to plan and implement the following:

- Developing the RSP.
- Executing the approved RSP.
- Use of available religious support resources.
- Allocation of assets.
- Location of UMTs.
- Coordination with indigenous religious leaders.
- Coordination with religious NGOs and IOs in coordination with Civil Military Operations leaders.
- Religious support for all authorized civilians in the AO.

PLANNING

3-3. The UMT participates in the collaborative planning process with the other staff sections known as the Military Decision-Making Process (MDMP). Religious support planning is continuous, detailed, and systematic. It examines all factors relating to the religious support of an operation. The planning process builds on the religious support section of the unit's tactical standard operating procedures (TSOP). It becomes specific to an operation through the preparation of a religious support estimate (RSE). The product of this planning process is a religious support plan (RSP). The RSP is included in the operations plan or operations order (OPLAN/OPORD). Inclusion may be in the plan/order as an appendix, annex or TAB. The specific placement of the RSP is dependent upon the particular unit's protocol for constructing operation plans and orders. The RSP informs the command and staff of the concept of religious support, and guides the work of the UMT. When appropriate, the RSP can be contained in a matrix format. A model for such a RSP matrix is contained in Appendix D of this manual. The use of a RSP matrix is appropriate when UMTs are assigned or attached to units that publish their OPLANS/OPORDS in a matrix format. The RSP matrix model is a self-explanatory form that clearly and concisely articulates how the UMT will provide religious support during the operation. The use of a RSP matrix is highly recommended when time constraints impact the planning process and the UMT needs a ready tool to direct and encapsulate its religious support planning.

KEY PLANNING CONSIDERATIONS:
- Commander's Intent
- Concept of Operations
- Concept of Logistics
- Concept of Medical Support
- ROE
- METT-TC of the battlespace/theater of operations

Religious Support Estimate

3-4. The estimate process is a problem solving procedure to assist the commander in determining the best course of action (COA) to accomplish a mission or task. Staff estimates are designed to supplement the commander's estimate within the respective areas of expertise. A staff estimate consists of significant facts, events, conclusions and recommendations on how best to use available resources to accomplish the mission. Upon receipt of the commander's guidance, each staff section develops its own staff estimate relating how its area of interest will affect the mission.

3-5. The chaplain develops the Religious Support Estimate (RSE) that becomes the basis for the Religious Support Plan (RSP). Development of the RSE ensures the careful and intentional planning of religious support. In the development of the RSE, the chaplain considers the —

- Location and movement of the UMT during all phases of the operation.
- Logistical and communication support required for specific missions.
- Replacement and/or augmentation of UMT members in the event UMT members become casualties or are unable to meet all RS needs during especially intense fighting.
- Religious, moral, and morale issues that affect the soldiers and the unit.
- Impact of indigenous religions on the military operation.

BASIC PRINCIPLES OF RS PLANNING AND COORDINATION

- Plan early and continuously.
- Use guidance from higher HQ.
- Exploit all available RS assets.
- Use the lowest echelon capable of furnishing effective RS.
- Use the most effective RS means.
- Furnish the type of RS requested.
- Avoid unnecessary duplication.
- Coordinate RS along command and technical channels.
- Provide adequate support.
- Provide rapid coordination.
- Provide safeguards and survivability.
- Establish RS coordination measures.
- Establish communications support.

3-6. The preparation of the estimate follows a logical, continuous, and methodical process to provide the appropriate religious support for the commander's selected COA. In conjunction with the other staff sections, the UMT prepares the RSE using the following steps:

1. Gather facts and assumptions.
2. Analyze the mission.
3. Develop courses of action.
4. Analyze and compare courses of action.
5. Make a recommendation.

Step 1: Gather Facts and Assumptions

3-7. The staff gathers information concerning the assigned mission and divides the information into two categories: facts and assumptions. Facts are statements of known data concerning the situation, and include the following:

- Enemy and friendly dispositions.
- Available troops.
- Unit strengths.
- Materiel readiness.
- Mission and the commander's intent.

3-8. Special religious needs of soldiers in the unit are key facts for the RSE. The team develops a Religious Preference Profile (RPP) using data from a Standard Installation/Division Personnel System (SIDPERS) ad hoc query. (See AR 680-29 for religious preference codes.) The report is a statistical analysis of the unit's religious demographics and provides the stated religious preference of soldiers assigned to that unit.

3-9. When facts essential to planning for RS are not available, the UMT must make assumptions. An assumption is a supposition about the current or future situation, assumed to be true, though without positive proof. In the absence of facts, assumptions are used to fill gaps. Assumptions are necessary only when they bear directly upon the planning or the execution of the plan. As new information becomes available, some assumptions become facts and some are modified or discarded.

3-10. Mission, Enemy, Terrain and Weather, Troops and support available, Time available, and Civil Considerations, or METT-TC, is a useful outline by which to consider and formulate facts and assumptions essential for developing an RSE.

Mission, Enemy, Terrain and Weather, Troops and support available, Time available, and Civil Considerations (METT-TC)

Mission: The UMT identifies specified or implied religious support tasks from the higher headquarters' OPLAN or OPORD. For example, a requirement to provide area religious support may be implied but not specified by a higher headquarters' task organization. Other considerations may include protection of religious shrines, artifacts or sacred places in the AO.

Enemy: In this part of mission analysis, the team identifies how the enemy may threaten the religious support mission. For example, enemy activity in rear areas may restrict team travel and necessitate additional area support requirements. The team may also consider how religion plays a part in the enemy's capacity or inclination to fight.

Terrain and Weather: The UMT assesses the terrain and weather in the AO. Traffic-ability, weather extremes, and the amount of daylight are factors that will affect the religious support mission. (Consider observation and fields of fire, cover and concealment, obstacles, key terrain, and avenues of approach (OCOKA).

Troops: Religious support is focused upon soldiers. What soldiers are doing often determines their availability for religious support. The UMT adjusts its plan to match the soldiers' schedules to meet their needs. The UMT must also assess the level of anxiety and fatigue among the troops to shape its ministry to the soldiers more effectively. The UMT will also advise the commander on the morale, moral, and spiritual health and welfare of soldiers.

Time Available: The UMT assesses the time required for planning and the time available for executing the religious support mission. The amount of planning time determines the detail of the plan. The team also considers the travel time to each unit, preparation time, and available light.

Civil Considerations: The UMT researches and advises the command on the religious attitudes and activities of the civilian population, religious and cultural dynamics, religious organizations, and religious leaders within the battlespace. Coordination with CMO personnel is essential. The chaplain advises the commander regarding civil concerns and their impact on the mission:

- Understanding the operational religious environment.
- Recognize religious implications for courses of action during planning.
- Anticipate and control negative impacts on the local religious climate.
- Consider long-term effects of operations on the indigenous religious climate.
- Ensure legitimacy and credibility of friendly forces.
- Balance operational priorities with concern for religious freedom.
- Minimize negative effects of operation on unit soldiers belonging to the same faith group as the local population.

Step 2: Analyze the Mission

3-11. Mission analysis begins with a review of the intent of commanders one and two levels higher. From this review, the commander and staff derive the specified tasks that the unit must perform to accomplish the mission. The unit must also identify implied tasks and determine which tasks are essential. A tentative list of essential tasks is presented to the commander for approval. Based upon the approved list, the staff presents the commander with a restated mission for approval. Commanders disseminate the restated mission as a warning order.

3-12. The UMT participates with the rest of the staff in mission analysis by identifying specified and implied religious support tasks. The UMT also considers any religious factors that may directly affect the mission. (FM 101-5, Appendix J)

3-13. During mission analysis, in addition to the RSE, the UMT brings expertise to the staff team in studying the following areas:

- ***Unit Morale.*** The UMT should have a sense of the morale of the unit. Factors such as whether the troops are fatigued, are excited or dreading the mission, and the spiritual health and welfare of the unit all impact on unit morale.
- ***Ethical and moral implications.*** The chaplain can assist the staff in considering the ethical and moral implications of a particular plan as it impacts on the lives of leaders, soldiers, and the local civilian population. For example, this may include compassionate treatment of EPWs, detainees, and displaced persons.
- ***Impact of local religion on the mission.*** The chaplain can assist in the religious assessment of an AO. The UMT should familiarize themselves with the local religions and their significance within the area of operations. What and where are the places that are considered sacred? What are the significant religious holidays and how are they celebrated? How could this impact the operation or the timing of the operation? (See Appendix F).

Step 3: Develop Courses of Action

3-14. Course of action (COA) development is based on mission analysis and on the facts and assumptions developed earlier. After receiving the commander's guidance, the Chief of Staff (CofS) or the executive officer (XO) leads the staff in developing COAs for accomplishing the mission. As the staff begins to develop COAs, the chaplain considers the religious and moral/ethical implications of each proposed COA and how to provide religious support. While the operations officer formulates the tactical possibilities, the chaplain and other staff officers consider how to integrate their functional areas into each COA. In the midst of COA development, the use of Chaplain Detachments (CD) by those units to which they have a doctrinal association must be considered and implemented into the COAs and the eventual RSP. Likewise, units with a doctrinal association with CDs must proactively request CDs through appropriate staff actions to ensure their operational employment.

Step 4: Analyze and Compare Courses of Action

3-15. After the staff develops several COAs, it analyzes each one. COA analysis consists of two parts: war-gaming and COA comparison. Through this process, the staff identifies the best COA to recommend to the commander.

3-16. Note that the chaplain may have a different kind of input depending on the kind of mission that is being planned. For units preparing for offensive or defensive actions, often the chaplain's role is more involved with integrating with the staff and being a team player while developing the RS aspect of the plan. In a stability and support operation, a chaplain may have more to contribute in the plan development process. In both cases, the chaplain will want to anticipate the COA toward which the commander may be leaning.

3-17. Additionally, the chaplain will want to pay particular attention to the casualty flow estimates and concepts of support. During the war gaming phase of the MDMP, chaplains must pay particular attention to where the casualties are expected, remembering that this critical ministry is one of the three major functions of comprehensive religious support. If the chaplain participates in the war gaming phase of the analysis, the chaplain will have a much clearer understanding of the scope of the mission when the COA is finally decided upon and then rehearsed.

Step 5: Make a Recommendation

3-18. After choosing the best COA, the staff briefs the commander. The commander may agree with the staff's recommendation, modify it, or select another. Once the commander decides on a COA, the UMT uses the concepts of support developed during the RSE process to prepare the Religious Support Plan (RSP).

3-19. The commander and staff begin preparations for the mission rehearsals. The chaplain needs to participate in the rehearsals in order to test the effectiveness of the RSP. Afterwards an assessment is made as to whether the plan actually meets the needs of the troops according to the commander's guidance.

Note: The adjutant (S1), supply officer (S4), and the chaplain are often not included in the accelerated MDMP process. However, this does not preclude the UMT from preparing RSEs and RSPs. Staff coordination and proper analysis is always necessary.

Religious Support Plan

3-20. Based on the selected COA, the UMT completes the details of its plan for inclusion in the OPLAN/OPORD. The religious support plan (RSP) assigns religious support responsibilities, defines area support requirements, and authorizes coordination between subordinate units.

3-21. Additional information found in the RSP may include the following:

- Concept of support for special faith groups and special holy days.
- Concepts of support for joint and multinational religious services support.

- Religious support requirements of all authorized civilians.
- Religious support for EPWs and Detainees in accordance with the Geneva Convention.
- Command guidance regarding liaison responsibilities with faith-based NGO/IOs and liaison support, in coordination with CMO personnel, to religious leaders in the AO.

3-22. Priority of religious support is—

- Unit military personnel.
- Other military personnel in AO.
- Authorized civilians such as DOD employees and contractors.
- EPWs and Detainees.

Planning Religious Support in Joint Operations

3-23. Just as in the Army planning process, the UMT is included in the system of joint planning. The Religious Ministry Support Team (RMST) (chaplain and chaplain assistants as designated under JP 1-05, which the Army designates as UMT) at each echelon is intricately involved with the orders process. Planning religious support operations must encompass and consider—

- Denominational assets.
- Command relationships.
- Location of RMSTs.
- Logistical requirements.

(See JP 1-05 for more detail.)

Planning Religious Support in Multinational Operations

3-24. Religious support planning with multinational forces in an operation must include the differences in religious support doctrine (or military custom of providing religious support), capabilities, force structure, logistics, and culture (denominational representation, language, supervision, etc). Communication, cooperation, and coordination are essential in planning multinational religious support operations. Senior force RMSTs need to understand the capabilities and limitations of religious support with each country represented within the coalition. Planning must remain flexible due to these differences while emphasizing the similarities that may include, language common between countries, denominational backgrounds of chaplains, and locations for area support. Senior U.S. RMSTs must be highly sensitive to allocating religious support assets of another country and should only do so with prior consultation and coordination. Planning considerations include—

- Communication requirements and limitations.
- Denominational representation.
- Language difficulties.
- Limitations in formal chaplain structures.

Rules of Engagement

3-25. Rules of engagement (ROE) specify the circumstances and limitations in which forces may engage the "enemy". Many factors influence ROE, including national command policy, mission, operational environment, commander's intent and the law of land warfare. Properly written ROE are clear and tailored to the situation.

3-26. Chaplains at all echelons help commanders apply the ROE by advising on the moral/ethical implications of proposed COAs. Senior-level chaplains consider the restrictions and constraints of the ROE when recommending religious support policy to the commander.

PREPARATION

3-27. Preparation is the link between planning and execution. While the commander and staff are planning, the unit prepares for the operation based on a warning order issued earlier by the commander. Once the plan is complete, preparations unique to the mission may begin. Upon completion of the individual and team preparations, the UMT provides religious support to soldiers while they complete their preparation for the operation.

Rehearsals

3-28. Units conduct rehearsals during preparation for combat. In addition to maneuver rehearsals, units conduct combat service support (CSS) and casualty evacuation rehearsals. Rehearsals provide an opportunity to refine the RSP. The UMT must be present at the rehearsal, and integrate the RSP into it, to fully understand the scheme of maneuver, support, and other critical pieces of information shared. If time permits, UMTs may conduct their own RS rehearsals following those of the unit. Such RS rehearsals offer UMTs an opportunity to resolve communication issues, as well as, build a comprehensive understanding of each UMT's religious support mission.

Individual and Team Preparation

3-29. The chaplain and chaplain assistant prepare themselves for combat like other soldiers of the unit. This preparation must be physical, mental and spiritual. The chaplain and chaplain assistant must also insure that their families are ready for coping with the challenges of deployment.

3-30. Team preparation includes load plans, pre-combat checks, updating information, map reconnaissance, travel planning, rehearsals, and mission coordination. Other preparations include securing equipment for the operation such as worship aids, devotional literature, music, and liturgical items.

Pre-Combat Inspections and Preparation

3-31. The UMT conducts a pre-combat inspection according to the unit TSOP. This inspection ensures the UMT possesses sufficient supplies and equipment to successfully execute the operation. On every mission, the team must carry food, water, and other survival equipment. The chaplain assistant prepares and updates the team's load plan and inventories and arranges the necessary supplies and equipment for movement.

Information Update

3-32. The UMT updates mission information regularly— especially before departing from its operational base. The tactical situation can change rapidly on the battlefield. The team checks the situation map, situation reports (SITREPS), logistics reports, and the latest fragmentary orders (FRAGOs) or warning orders. If there are significant changes in the situation, the team modifies its plan accordingly.

Training

"The key to winning on the battlefield is the understanding of "how we fight" and the demonstrated confidence, competence, and initiative of our soldiers and leaders. Training is the means to achieve the tactical and technical proficiency that soldiers, leaders, and units must have to enable them to accomplish their missions."

FM 7-0, TRAINING THE FORCE

3-33. Training is the cornerstone of preparation for military operations. The Army's primary activity during peacetime is realistic, battle-focused training oriented on the unit's mission essential task list (METL). Installation chaplains, through their training managers, and in coordination with subordinate UMTs and unit operations/training officers (S3/G3), ensure that UMTs have adequate technical and tactical training for deployment and combat. The training objective is operational readiness—units and soldiers that are tactically and technically proficient.

3-34. Tactical proficiency is a demonstrated understanding of the Army's war fighting doctrine and tactics. To provide effective religious support in combat, the UMT must know its unit's standard operational procedures, tactics, and techniques. The UMT must also maintain the requisite skills of field craft essential for survival on the battlefield. Tactical training is primarily the responsibility of the unit to which the UMT is assigned or attached.

3-35. Technical proficiency is the demonstrated ability to perform all tasks related to one's functional area. Technical proficiency is required for both garrison and field missions. Technical proficiency is acquired from schools, self-study and training with other UMTs in performing the unit's and installation's religious support mission. Training and continuing education conferences, functional courses offered by the U.S. Army Chaplain Center and School (USACHCS), and courses offered through the Officer and NCO Education System (OES and NCOES) supplement the installation religious support-training program.

"To understand firepower and its application you have to use it, actually practice with it and see it work. This means maneuvers in the field, maneuvers that reproduce as realistically as possible, within the bounds of prudence, the actual conditions a soldier will encounter on the battlefield. Only so can you condition, not only the human body, but also the human spirit, to face and survive the stresses and strains the soldier will encounter in battle. And in future battle, if it comes, these stresses and strains will be immeasurably greater."

General Matthew Ridgeway

3-36. The UMT must plan for and participate in religious support training for both combat and garrison environments. Furthermore, the UMT must participate fully in both the unit tactical and religious support training to assure their tactical and technical proficiency.

Assessment of Unit Ministry Team Readiness

3-37. As the commander's personal staff officer for religious support, chaplains at brigade and higher levels are responsible for assessing the tactical and technical proficiency of UMTs in subordinate units. They conduct assessments of the UMT during field training exercises (FTXs), external Army Training and Evaluation Program (ARTEP) events, unit training and Command Inspections. Observer controllers provide additional assessment assistance when the UMT is involved in Battle Command Training Program (BCTP) events or Combat Training Center (CTC) rotations. These CTC assessments for UMTs are important for readiness analysis and training.

3-38. Chaplains and chaplain assistants participate in the installation's Command Master Religious Plan (CMRP) to help soldiers develop the personal spiritual readiness needed to sustain them during combat. Spiritual readiness is also essential to sustain soldier families while the unit is deployed. As part of the installation's CMRP, the UMTs conduct worship services, offer religious education, and provide spiritual fitness training and other activities in installation chapels and facilities, and in the theater of operations. Spiritual readiness applies to all levels of the command. Planning for spiritual readiness programs must be part of the preparation for combat for each soldier and authorized personnel.

Religious Support Preparation in Joint Operations

3-39. Coordination between all RMSTs is critical for full implementation of the RSP. Total synchronization between all religious support elements is key to a comprehensive plan. The joint task force (JTF) RMST considers the following additional tasks for operational preparation:

- Identifying RMST logistical requirements.
- Personnel staffing and augmentation requirements.
- RSP integration within the operational plan.
- RSP rehearsals.
- Spiritual Nurture programs for the command and staff.
- Loading of equipment and material.

3-40. The Command JTF RMST works closely with the Joint staff for RMST closure within the Joint Operational Area (JOA), ensuring communication and synchronization of the RSP.

Religious Support Preparation in Multinational Operations

3-41. Each nation brings different capabilities and limitations to the Combined-Joint Task Force (CJTF). Task organizing along functional lines vice national lines can help fill capability gaps, and designs each nation's role so they are full participants. Comprehensive religious support information disseminated to all participating nations describing religious support operational expectations will prepare all combined participants for spiritual and religious coverage. UMTs must logistically prepare themselves to meet the many diverse and varied religious and cultural differences. Communication limitations and differences must be prepared for such as language, terminology, systems, and distance. Liaison and training can offset some of the challenges.

"A chaplain visits our company. In a tired voice, he prays for the strength of our arms and for the souls of the men who are to die. We do not consider his denomination. Helmets come off. Catholics, Jews, and Protestants bow their heads and finger their weapons. It is front-line religion: God and the Garand."

Audie Murphy, To Hell and Back

COORDINATION

3-42. Coordination consists of three categories: mission, internal, and external.

Mission Coordination

3-43. The size of the AO and unit dispersion challenge the UMT's ability to coordinate for religious support. Communications equipment is critical for timely and effective coordination. Also, key communications systems allow the UMT to be available for emergency tasks. Early planning and coordination will help to eliminate potential problems. Coordination is both internal to the participating units and UMTs, and with external units, agencies, and UMTs.

Internal Coordination

3-44. As a personal staff officer, the chaplain works for the commander, advises the commander, and executes the commander's religious support program on his behalf. The UMT coordinates the execution of the commander's religious support program with the Chief of Staff or XO. The team plans cooperatively with all staff sections for the exchange of information.

External Coordination

3-45. Because of the rapidly changing dynamics, conditions, and circumstances on the modern battlefield, coordination for area support is difficult. Although religious support execution is decentralized, UMTs must synchronize religious support within their subordinate units. Establishing synchronization procedures with subordinate commands before deployment reduces the communication and coordination challenge for the UMT.

3-46. Military communication channels streamline information distribution on the battlefield. The team uses command, staff, and technical channels for communication, guidance, and coordination.

- ***Command:*** The command channel is the official communication line between headquarters for passing orders, instructions, and tasks to subordinate units. Command channels link commander to commander. Within the authority granted them, staff officers use command channels when acting for the commander. It is seldom if ever used by the UMT.
- ***Staff:*** The staff channel is the staff-to-staff link between headquarters. It is the primary channel the UMT uses for planning and coordinating religious support. Through this channel, the team transmits and coordinates planning information and provides staff supervision of religious support in subordinate units.
- ***Technical:*** The technical channel is the link between members of the same branch or discipline to convey technical instructions and guidance. Chaplains use the technical channel to communicate with UMTs of other

units about the technical aspects and details of religious support. It is not used to send official reports, to request religious support, or to send routine information between teams. This information is communicated through command or staff channels.

EXECUTION

3-47. The execution of the religious support mission on the battlefield is decentralized and performed by UMTs in accordance with the RSP. Adjustments to the plan are coordinated between affected units.

RELIGIOUS SUPPORT PLANNING FOR FORCE PROJECTION

MOBILIZATION

3-48. United States military strategy rests upon the ability to rapidly respond to crises. Responding to crises requires the Army to deploy its forces and may require mobilization of Reserve Component (RC) resources. Deployment comprises those activities required to prepare and move the force and its sustainment equipment and supplies to the AO.

Religious Support during Mobilization

3-49. Mobilization is the activation of the U.S. Army Reserve (USAR) and federalization of the Army National Guard. Mobilization provides commanders with RC forces, manpower, facilities, and logistics. FM 100-17-3, Reception, Staging, Onward Movement, and Integration (RSOI), provides additional information on RSOI procedures. It expands the Army's capability to respond to crises. Mobilization is conducted in five phases:

- Phase I. Planning.
- Phase II. Alert.
- Phase III. Home Station.
- Phase IV. Mobilization Station.
- Phase V. Port of Embarkation.

Mobilization Phase I: Planning

3-50. Planning begins long before a unit is mobilized. During peacetime, Active component (AC) and Reserve component (RC) UMTs coordinate with AC units for their mobilization planning. This planning takes into account operational and training requirements, equipment status, and readiness. It also includes the impact of mobilization on soldiers, families, and communities.

3-51. The primary planning tool for mobilization is the Unit Ministry Team Mobilization Systems Planbook (MOB Planbook). The MOB Planbook integrates all phases and levels of mobilization, and it provides a detailed guide for integrating mobilization planning.

3-52. Installation, Continental United States Army (CONUSA), STARC, and USARC chaplains are key to Reserve Component UMT mobilization. Each maintains a copy of the UMT MOB Planbook. Organizational MOB Planbooks contain the same basic information, but each is tailored for the particular headquarters. The MOB Planbook specifies what must be accomplished during each phase of mobilization. While many of the tasks in the

MOB Planbooks are unique, they are designed to integrate the mobilization process.

3-53. A unit's success is directly related to the quality of its preparation before it conducts operations. A representative listing of the mobilization planning responsibilities for installation, CONUSA, STARC, and USARC chaplains is detailed in Chapter 7. Planning considerations—

- Does the home station require special planning or preparation?
- Where is the home station in relation to support resources?
- How does the UMT provide family support briefings?
- How will the UMT provide required religious support if some parts of the unit remain at home station for an extended period of time, while others elements are sent forward to the mobilization station (MS)?
- Does the UMT go forward with the advance party or with the main body of the unit?
- How do durations in personnel holding areas, movement, or separation influence soldier morale?
- What religious support activities would address morale problems in the unit?
- What actions are required to close out civilian responsibilities?
- What coordination is required with faith groups for religious coverage during mobilization?
- What supplies are required for religious support?
- What religious support is required as the unit transitions to war?
- What training is required to bring the UMT to validated status for deployment?

3-54. Thorough planning and realistic training before a mobilization reduces confusion between all key UMT participants. Mobilization training gives both AC and RC UMTs an opportunity to practice their mobilization religious support missions, and an opportunity to assess the adequacy of mobilization plans. Installations assist with the following RC UMT training events:

- Annual training (AT).
- Individual Mobilization Augmentee (IMA) training.
- Individual training events, Individual Ready Reserve (IRR), and Chaplain Candidate training.
- AC/RC Sustainment Training.
- Training for UMTs in Garrison Support Units (GSUs).
- AC / RC Annual Chaplaincy Regional Sustainment Training (CREST) as mandated by the Chief of Chaplains.

Planning Considerations:
• Does the home station require special planning or preparation? • Where is the home station in relation to support resources? • How does the UMT provide family support briefings? • How will the UMT provide required religious support if some parts of the unit remain at home station for an extended period of time, while others elements are sent forward to the mobilization station (MS)? • Does the UMT go forward with the advance party or with the main body of the unit? • How do durations in personnel holding areas, movement, or separation influence soldier morale? • What religious support activities would address morale problems in the unit? • What actions are required to close out civilian responsibilities? • What coordination is required with faith groups for religious coverage during mobilization? • What supplies are required for religious support? • What religious support is required as the unit transitions to war? • What training is required to bring the UMT to validated status for deployment?

Mobilization Phase II: Alert

3-55. The Alert Phase begins when the unit receives the alert order of a pending mobilization order. The UMT reports immediately to its home station and the unit conducts the final screening of soldiers and cross leveling of personnel and equipment. The Alert Phase ends on the effective date of the unit's entry on active duty.

Note: RC chaplains cannot be mobilized (i.e. cannot be brought onto Active Duty) until they have completed all phases of the Chaplain Officer Basic Course (CHOBC). AC chaplains are non-deployable until the sustainment phase of CHOBC is completed. (See AR 165-1)

Mobilization Phase III: Home Station

3-56. Phase III begins on the effective date of the unit's mobilization. The effective date of mobilization is when RC chaplains and chaplain assistants come on Active Duty. During this phase the unit begins its transition to active duty. At home station the unit inventories property, dispatches an advance party, and moves to the mobilization station (MS). This phase ends when the unit arrives at the MS.

3-57. UMTs must balance the time spent in their own deployment preparation with providing religious support to soldiers and families.

Mobilization Phase IV: Mobilization Station (MS)

3-58. Phase IV begins when the unit arrives at the mobilization station. Activities at the MS include processing personnel and equipment, accessioning the unit into the active structure, cross-leveling, and soldier readiness processing. The UMT, along with the unit, engages in individual and collective training and validation exercises for survival on the battlefield. The goal of the unit during this phase is to achieve mission capability in the shortest possible time. The MS phase ends when the unit arrives at the port of embarkation (POE).

Note: The unit's line of authority shifts during this phase from the CONUSA to the MS commander.

Mobilization Phase V: Port of Embarkation (POE)

3-59. This phase begins with the arrival of the unit at its POE. Actions at the POE include preparing and loading equipment and manifesting and loading personnel. The POE phase ends with departure of personnel and equipment from the POE.

UNIT RELIGIOUS SUPPORT DURING DEPLOYMENT

3-60. Deployment is the planning, preparation and movement of forces and support base from any location to an area of operations in response to military need or crisis. Deployment may take the form of a forcible or permissive entry. Deployments are conducted in five phases:

- Phase I. Pre-deployment Activities.
- Phase II. Movement to Ports of Embarkation.
- Phase III. Strategic Lift.
- Phase IV. Theater Base Reception.
- Phase V. Theater Onward Movement.

Deployment Phase I: Pre-deployment Activities

3-61. Pre-deployment activities begin when a unit receives a deployment notification. These activities include mission briefings and planning, family pre-deployment briefings, equipment loading, chapel closeout, and the transfer of property accountability.

Mission Planning

3-62. As discussed above, the commander and staff begin the MDMP after they receive the deployment notification and they review contingency plans for completeness and accuracy. The UMT, along with the rest of the staff, begins to prepare an estimate of the situation. The team balances planning and preparation with providing religious support.

3-63. Religious support planners at higher echelons integrate the broader scheme of the operation into the planning process. They assess lines of authority, task organization, and faith group needs to determine the required number and mix of UMTs. Senior planners monitor the tailoring of the deploying force. When a JTF or Army Forces (ARFOR) commander forms an ad hoc headquarters, senior chaplains must monitor its composition to ensure the inclusion of a senior chaplain on the staff.

3-64. If the theater chaplain is not scheduled to deploy immediately, he designates a senior chaplain to be responsible for coordinating religious support in theater during each stage of the deployment preceding his arrival.

Chapel Transitions

3-65. The installation chaplain is responsible for all religious support facilities on the installation. Deploying UMTs coordinate chapel transition with the installation chaplain who decides whether to continue, combine, or suspend specific chapel programs and services. The UMTs transfer property accountability to the installation chaplain before departure. Deploying UMTs, in concert with non-deploying UMTs, provide transition activities for soldiers and their families.

Load Lists and Plans

3-66. The UMT must determine what to ship and what to carry. Some religious support supplies (communion bread, wine or grape juice) are consumable and cannot be pre-positioned with other war stocks. Sacred scriptures, religious literature, rosaries, crosses and medals can be pre-positioned. Equipment and supplies may not be available in theater for an unforeseen length of time. The UMT must include in the TSOP a religious support supply load list and plan.

Family Support Activities

3-67. As information about the deployment becomes available, the unit, as directed by the commander provides pre-deployment briefings to families (typically the S1 coordinates such briefings). The information given will be constrained by operational security requirements. Pre-deployment briefings include information on services that will be available on the installation during the deployment. Information will also be discussed regarding the advantages or disadvantages of family members remaining in the military community or returning to their community of origin. UMTs participate in pre-deployment briefings to provide religious support information to families.

3-68. Religious support provided to soldiers and their families directly contributes to the success of mobilization and deployment. Religious support fosters family wellness, a substantial factor in soldier readiness. Lessons learned from previous deployments and other research establish that soldiers deploying while anxious about personal and family problems are more vulnerable to combat stress. They are more likely to experience panic, poor judgment, battle fatigue, and the loss of the will to fight.

3-69. A lack of sufficient information, concern for coping without a spouse or parent, and the trauma brought on by a soldier's departure can generate stress within the family. Because families need religious support during this period, many will turn to the UMT. The deploying team must balance the needs of soldiers and families with its own preparation for departure. The non-deploying UMTs must assume much of the religious support responsibility for the families of the deploying soldiers.

3-70. It must also be mentioned that while care for the families of fellow soldiers is a responsibility for the UMT, the families of UMTs will also need care and religious support. Non-deploying UMTs can assist these families through the deployment process.

Deployment Phase II: Movement to the Port of Embarkation

3-71. When pre-deployment activities are concluded, the unit begins movement to the port. Units traveling outside of CONUS use both sea and aerial ports of embarkation (SPOE/APOE). Heavy equipment and the soldier operators travel by ship to the theater of operations. The remainder of the unit usually travels by air. Splitting the unit presents a challenge for the UMT.

Since the UMT will normally travel by air, RS for soldiers traveling by ship will require special planning and coordination.

Holding Areas

3-72. While the unit awaits transportation to the SPOE/APOE, it is normally confined to a Personnel Holding Area (PHA). Time spent there can seem long and monotonous. If the holding area is on the installation, this can be a frustrating time for soldiers because families are near but out of reach. This can be an important opportunity to provide religious support.

Deployment Phase III: Strategic Lift

3-73. Strategic lift begins with departure from the POE and ends with the arrival in theater. Transportation Component Commands are responsible for the strategic transportation of forces and their support. Religious support during this phase is oriented toward soldiers traveling with UMTs to the theater of operations.

Deployment Phase IV: Theater Base Reception

3-74. Force reception begins when the unit arrives at the port of debarkation (POD) in theater and ends when the unit departs the POD to begin the onward movement phase. One of the primary requirements during this phase is coordinating the onward movement of the deploying forces to their destination. The theater chaplain is responsible for religious support in the ports of debarkation.

3-75. Except in the case of opposed entry, CSS units may arrive early in the deployment to process combat units through the POD and to establish logistical support systems. Religious support resources in the port-based CSS units are limited. To prepare for operations, units are moved out of the PODs as rapidly as possible, through marshaling or staging areas, and into assembly areas (AA) or logistics bases.

3-76. In assembly areas soldiers prepare for onward movement. The unit may task organize the force, develop and issue tactical plans and orders, coordinate with other units, conduct reconnaissance, conduct training, and rehearsals. The immediacy of the pending mission frequently focuses soldiers' concerns upon matters of faith or religion. When units are in Assembly Areas (AA), UMTs are generally able to provide religious support to soldiers with little risk and with great receptivity.

Deployment Phase V: Theater Onward Movement

3-77. Onward movement begins with the personnel and equipment linkup, the reconfiguration of forces, sustainment, and receipt of pre-positioned war reserve stocks at designated areas. This phase concludes with the unit's arrival at the gaining command's Intermediate Staging Base (ISB) and tactical assembly areas.

Intermediate Staging Base

3-78. Units traveling through the ISB en-route to their combat mission present some religious support challenges. ISBs vary in size depending on the forces involved in the mission, and at times can be large in terms of population and geography. Units remain in the ISB for a minimum time, thus creating a highly transient community.

Tactical Assembly Areas

3-79. Tactical assembly areas (TAAs) are occupied by brigades, battalions, and companies that are forward of the rear area. Enemy contact is likely and the commitment of the unit directly into combat is possible and often anticipated.

3-80. Actions in the TAA focus on preparing the unit for future operations. Common TAA activities include task organization, maintenance, personnel replacement, re-supply, reorganization, rest, and planning for future operations. UMTs monitor changes in the task organization and adjust their plans accordingly. TAAs offer good opportunities for religious services.

Continental United States Replacement Centers

3-81. Replacement centers within the Continental United States (CONUS) receive and certify individuals for deployment to a combat theater. The CONUS Replacement Centers (CRCs) process non-unit related AC soldiers (to include RC soldiers accessioned onto active duty), Department of the Army civilians, contract civilians, Red Cross workers and other civilians.

BATTLEFIELD TRANSITION PLANNING AND EXECUTION

3-82. As with planning for different types of operations, planning for mission transitions must extend throughout the planning process and into operations and redeployment. Transferring control of an operation to another organization requires detail planning and execution. Transition depends upon the situation and will have its own characteristics and requirements. The UMTs should continuously communicate and coordinate among themselves, ensuring continual RS. After a hostile conflict or engagement, regeneration of force capabilities will be a primary consideration for the UMT in the transition plan. Following a stability or support operations mission, the soldier's physical, emotional, family, and spiritual recovery may be the UMT's transition focus. Mission analysis should provide the initial information to commence transition planning. The RSP supports the command's OPLAN with thorough integration. Transition planning and execution considerations include—

- Logistical supplies.
- Synchronized RSP transition with follow-on units.
- Denominational support.
- Transition religious support continuity file.
- Religious and Cultural information of the AO.

3-83. A smooth transition between forces and UMTs is essential for continuous religious support of all the military units. Cooperation among UMTs will ensure that all military and approved personnel receive quality religious support.

RELIGIOUS SUPPORT DURING REDEPLOYMENT AND DEMOBILIZATION

3-84. With the cessation of hostilities or the declaration of a truce, the deployed force in a theater of operations will transition through the last stages of Force Projection. This transition from the post-conflict stage through the redeployment and demobilization stages may begin even if residual combat operations are still underway.

3-85. While some forces are preparing for redeployment, other forces remain in support of Peace Building Operations. While soldiers restore order and reestablish the host nation infrastructure, UMTs provide religious support during the post-conflict stage.

3-86. As operations cease and forces are no longer required for Peace Building, units begin the redeployment stage. Forces and materiel not required for subsequent operations re-deploy to CONUS or home-theater to prepare for future missions. Redeploying forces may deploy to another theater or to an area other than home station on short notice. Redeployment, like deployment, occurs in several phases:

- Reconstitution for strategic movement.
- Movement to redeployment assembly areas.
- Movement to the port of embarkation.
- Strategic lift.
- Reception at a port of debarkation.
- Onward movement from port of debarkation.

3-87. During demobilization the process of transferring RC units, individuals, and materiel from an active posture to a pre-mobilization or other approved posture begins. Demobilization coincides with the first three phases of redeployment, the phases before the units depart the theater of operations. During this stage, RC units return to pre-mobilization status.

3-88. Demobilization requires careful planning which takes into account public relations and what the soldiers are thinking and saying to each other.

3-89. The phases of demobilization are—

- Planning.
- Area of operations demobilization actions.
- Port of embarkation to demobilization station operations.
- Demobilization station operations.
- Home station actions.

Unit Religious Support During Redeployment

3-90. Religious support during redeployment is similar to religious support during deployment. Some soldiers and contractors will be preparing equipment for strategic lift or waiting in holding areas for transportation, while others will be engaged in force projection operations.

> Worship, pastoral care, and religious education help soldiers begin the transition back to normal life.

3-91. UMTs providing religious support in assembly and PHAs during redeployment face challenges similar to those encountered during deployment. Worship, pastoral care, and religious education will help soldiers begin the transition back to normal life. Reunion sessions, an important part of the UMT's work, require early planning to be effective.

Reunion

3-92. Reunion denotes the reuniting of family members after separation as a result of deployment. Army research shows that family problems intensify when spouses return from deployments. Characteristics of the readjustment period, which lasts for about two months, include the emergence of physical symptoms, children's behavior problems, and marital tensions. For many couples, the first weeks of reunion are the most challenging part of the whole deployment experience. Tensions become most acute during the two weeks before and after the soldiers' return.

3-93. Religious support during reunion assists soldiers and family members in dealing with the stress that may occur while attempting to integrate experiences while separated, as well as, personal expectations of family life after deployment. UMTs help to restore and preserve the relationships that couples and families knew before deployment. During the reunion, UMTs join with the Installation Religious Team, Army Community Services, STARCs, United States Army Reserve Command (USARC) Major Subordinate Commands (MSC), family program coordinators, and mental health personnel to provide family assistance.

> By assisting with the stress of reunion, the UMT helps to restore and preserve the relationships that couples and families knew before deployment.

3-94. As married soldiers encounter reunion issues with spouses, single soldiers will experience reunion struggles with family and friends. Soldiers deployed as individuals or in small detachments also experience reunion problems upon return to their units. Reunion briefings and workshops help to minimize the problems for soldiers, families, and units. These events equip soldiers with coping skills to face the challenge of reintegration to families, peer groups, and units.

Reunion Preparation In Theater Before Redeployment

3-95. Before departing a theater of operations, UMTs coordinate with the S1/G1 for reunion briefings and workshops. The workshops teach skills needed to reconnect with younger children or to bond with newborns. During the pre-reunion period soldiers and families wonder, "What will it be like to be together again?" and "How has this experience changed us and our relationship?" While deployed UMTs conduct workshops in theater, non-deployed UMTs conduct similar workshops for waiting families.

> UMTs coordinate with the S1/G1 for briefings and workshops for soldiers in order to prepare them for reunion.

Reunion Preparation at Home Station after Redeployment

3-96. UMTs plan additional workshops at home station. Some workshops will be for soldiers and spouses, others for soldiers or spouses alone. UMTs may coordinate workshops with the Installation Ministry Team (IMT), ACS, Family Readiness Groups (FRG), and Chaplain Family Life Centers. The UMT must be prepared to sustain the effort for several months, depending on the length of the separation and the severity of the mission. From time to time, UMTs conduct assessments of the effectiveness of the workshops and the need to continue them.

3-97. Members of UMTs are cautioned not to ignore their own family reunion process. Being occupied with helping others is not a reason for neglecting one's own family. Installation and Family Life Chaplains conduct workshops specifically for redeploying UMTs.

> UMTs are cautioned not to ignore their own family reunion process.

Recovery/Refitting

3-98. The unit's need to reestablish its readiness level sometimes competes with the soldier's family for the soldier's time, energy, and attention. During the recovery and refitting phase, soldiers need to clean, repair, and secure equipment; identify shortages; requisition supplies; and reestablish stock levels. Property accountability is reestablished with the installation Property Book Officer (PBO). UMTs are not exempt from these concerns, even though they are busy helping soldiers and families to reintegrate.

> UMTs must balance reunion activities with refitting.

RELIGIOUS SUPPORT DURING DEMOBILIZATION

Planning Actions

3-99. Demobilization planning begins with mobilization and ends with the decision to release units and individuals from active duty. This planning includes—

- Relocating units and individuals.
- Incorporating annual training (AT) for non-mobilized RC units to support demobilization.
- Mobilizing additional RC units.
- Using individual volunteers (by means of temporary tours of active duty) to support critical missions in the demobilization process.
- Basic maintenance of equipment.
- Awards and decorations.
- Evaluation reports.
- After Action Reviews/Lessons Learned.
- Chaplaincy Detachment (CD) and GSU integration for support.

In Theater Activities

3-100. Before a unit departs the theater, it initiates demobilization administrative and logistical actions to include—

- Processing awards, decorations, and performance reports.
- Initiating line-of-duty investigations.
- Updating supply records and equipment accountability.

3-101. Medical and dental actions are started in theater when resources are available. During this phase soldiers may be frustrated as they become anxious to return home. As always, UMTs adapt religious support based on an assessment of soldiers' needs.

Home Station Activities

3-102. The reintegration of soldiers who deployed with those who did not is an important consideration for UMTs. During mobilization, many soldiers may deploy at different times, or with other units as a result of cross leveling. During demobilization, they may also return to home station at different times. Feelings of resentment may build in the unit, and in families, because some soldiers have returned and others have not. Reentry into the civilian work force is of another concern of the soldiers of the RC.

INSTALLATION RELIGIOUS SUPPORT

REINTEGRATION/REUNION

3-103. The Installation Religious Team plans for—

- The reintegration of returning soldiers with their family members and with non-deployed soldiers in their units.
- Support to casualty survivors.
- Non-deploying UMTs to continue their coverage of units and family members normally served by the returning UMTs, so that they will be able to reintegrate without the pressure of immediate duty.

3-104. Chaplain Family Life Center will—

- Assist in preparing families for reunion with soldiers.
- Assist UMTs with training and technical guidance on reunions.

(See Chapter 7 for further discussion of the Installation Chaplain and NCOIC responsibilities.)

GENERAL RESPONSIBILITIES OF THE UNIT MINISTRY TEAM DURING RELIGIOUS SUPPORT OPERATIONS

Staff Supervision

3-105. Chaplains at brigade and higher levels provide staff supervision of religious support by visiting subordinate units, monitoring the tactical situation, and revising estimates and plans based on changes in METT-TC.

3-106. A staff officer does not have the authority to order a subordinate unit to execute directives or orders. Staff officers make recommendations to subordinate commanders. These recommendations may be accepted or rejected. A staff chaplain may contact a subordinate UMT to provide instructions, advice, and recommendations, offer assistance, or exchange information. The commander will establish the nature and degree of the chaplain's actual authority or control in implementing the RSP.

Advising

3-107. Chaplains advise the commander on religion, morals, ethics, and morale as affected by religion. Advice extends to concerns such as the quality of life for soldiers and families, impact of indigenous religions on operations, and the allocation of resources. Commanders rely on the advice of the UMT to help them discern the unit climate. At brigade and higher levels, the UMT advises the command on the assignment of chaplains and chaplain assistants. The chaplain's direct access to the commander allows face-to-face communication and is the most effective means for providing advice.

Administration

3-108. Administrative actions continue in the tactical environment. They comply with unit TSOPs and appropriate regulations and doctrine.

Correspondence

3-109. Battlefield conditions complicate the preparation of correspondence. The necessity for correspondence, however, will not decrease. The UMT must prepare required reports and correspondence despite equipment limitations or the tactical situation. Prior to deployment, UMTs must consider how to adapt their procedures to field conditions.

Records Management

3-110. Army regulations and the unit TSOP determine the procedures for managing and maintaining records. The UMT may need to maintain the following records:

- Baptisms and other sacramental acts.
- Services of worship and attendance.
- Ministration to casualties.
- Memorial services or ceremonies.
- Staff journal (See DA Form 1594).
- After Action Reviews.

Logistics

3-111. Logistic support is essential to religious support in combat. Because support requirements are more difficult to meet in combat, UMTs must become familiar with the military supply system and use it as their first source of re-supply rather than making local purchases. UMTs should know what items are available within the system, which items are authorized for the UMT, and what procedures to use to requisition supplies.

3-112. The Modified Table of Organization and Equipment (MTOE) is the authorization document for ordering equipment. The chaplain is responsible for all team equipment; while the chaplain assistant is accountable for its proper use, security, and operational readiness. Request for changes to the MTOE should be coordinated and staffed through the chain of command using DA Form 2028 (Recommended Changes To Publications and Blank Forms).

Supplies

3-113. The chaplain assistant monitors supply levels, conducts inventories, and requisitions supplies through unit supply channels digitally using Department of Defense on-line requisition sites, or in coordination with the Defense Supply Center Philadelphia (DSCP) Item Manager. This procedure will ensure timely replacement of standard expendable items. The team reports its supply requirements to the supply section by national stock number (NSN), nomenclature, amount, and projected usage rates.

3-114. UMTs should deploy with at least 60 days of ecclesiastical supplies. Normally the unit establishes stockage levels, expressed as "Days of Supply" (DOS). The UMT may need to adjust the amount of supplies carried in accordance with the projected rate of resupply for each mission. A team deploying to an immature theater with limited logistical support may not have resupply available for 90-120 days.

3-115. UMTs at division and higher echelons coordinate directly with the Defense Logistics Agency (DLA) Liaison Officer and the 9925 Item Manager at DSCP to ensure a steady flow of Chaplain Resupply Kits and other religious support supplies to subordinate teams.

3-116. Planning must incorporate religious education/sustainment supplies for long-term continual operations. These supply items are also managed by the 9925 Item Manager. The supplies will support all faith groups' spiritual sustainment needs.

Maintenance

3-117. Maintenance is critical in a combat environment. The team is responsible for the maintenance of its assigned vehicle and other equipment. An effective maintenance program is essential to ensure the UMT's ability to perform religious support. Operator maintenance focuses on preventive maintenance checks and services (PMCS). Both the chaplain and the chaplain assistant must be able to operate and maintain the vehicle. These skills must be acquired and practiced before deployment.

Transportation

3-118. Section 3547 of 10 USC mandates transportation assets be afforded to the Chaplain for the conducting of religious services. The vehicle must be able to transport the team to all unit locations within the AO. The team must take precautions to avoid compromising the unit's location or fighting positions. Both members of the UMT must be trained and licensed to operate the vehicle.

3-119. METT-TC will determine who drives the vehicle. The practice of the chaplain driving in hostile areas ensures that the assistant, who is a combatant, can provide security. When not providing security for the team, the chaplain assistant operates the vehicle.

3-120. The UMT should take full advantage of unit convoys. It must be thoroughly familiar with convoy procedures, unit policy regarding movements, and vehicle densities.

Communication

3-121. As the UMT travels on the battlefield, it must maintain communications with its operational base to keep the base informed of its status, location and travel plans. At the same time, the team receives and transmits situation reports (SITREP). The commander must provide the team with dedicated communications equipment that is compatible with and keeps pace with the commands secure communication equipment. The team may use a variety of other means available within the unit. The team must be flexible and creative to maintain communications.

3-122. Religious support is synchronized and implemented throughout the area of operations, theater of operation, and communications zone (COMMZ) by being fully integrated into the digitized communication and automation systems. The ability to reach from the area of operations to the home station support node (HSSN) is essential for relaying time sensitive information and religious support requests. A mobile communication link between each echelon enables the UMT to allow time-sensitive data to be rapidly communicated to and from each echelon. This is essential for the UMT's mission of providing soldiers with the personal delivery of religious support. It ensures that the UMT has the ability to move to where time-critical religious support is essential. The communication linkage between UMTs provides immediate access to processed critical information needed to ensure the commander's religious support plan is executed successfully.

Assessment

3-123. UMTs continuously identify, assess, and disseminate lessons learned throughout the force and branch. Assessment of lessons learned from operations is the key element for future training objectives. Observation teams deploy with Army forces on every major operation. Religious support is an important element of the operation and lessons learned must be gathered on and integrated into the overall assessment of the operation.

3-124. Lessons learned are also critical during a transition phase between one force and another. UMTs arriving into the theater of operations can better understand the battlefield and METT-TC implications through lessons gathered by departing UMTs.

RELIGION, MORALS, ETHICS, AND MORALE

3-125. Chaplains serve as the "conscience of the command." Chaplains advise the commander on the moral and ethical nature of command policies, programs and actions, as well as, on the impact of command policies on soldiers. They advise the commander on—

- Religious/ethical issues as they bear on mission accomplishment and morale.
- Inter-faith relations among religious groups in the command.
- The role of religion within indigenous population groups.
- Accommodation of special religious needs/requirements.
- Quality of life issues.
- Unit command climate.

INDIGENOUS RELIGIONS

3-126. The chaplain advises the commander and staff on the beliefs, practices, and customs of religious groups in the AO, and the impact that local religions may have on the conduct of the mission. This advice is based on information developed at division and higher echelons and is coordinated with the G5 and Civil Military Operations (CMO) officer. (See Appendices A and F for more information.) Common concerns include—

- Religious organizations and doctrines.
- Religious practices and customs.
- Places of worship, shrines, and other holy places.

PUBLICATIONS

3-127. Prior to deploying, UMTs review the list of publications in the reference section of this manual to determine what to carry. As a minimum, the team carries this manual (FM 1-05), JP 1-05, AR 165-1, and pertinent "go to war" publications of the unit. The chaplain assistant requisitions missing publications and maintains the team's publications.

JOURNALS AND WORKBOOKS

3-128. The UMT should maintain a staff journal covering each 24-hour period (See DA Form 1594). The staff journal is a chronological record of events and serves as a permanent record for operational reviews, training issues, and historical research. The team's journal entries should provide enough detail to describe its activities adequately and to fix the time and place of significant religious support events; however, care must be taken to avoid compromising privileged information by recording it in the staff journal. Journals are useful in reconstructing events, maintaining religious support histories, writing AARs and lessons learned for the operation. Typical entries include—

- Summaries of plans and orders.
- Religious support activities.
- Reports sent and received.
- Reports of contact with other UMTs, missionaries, faith-based NGO/IOs, and humanitarian agencies.

3-129. A UMT may choose to keep a staff section workbook, also known as a "battle book." It is an informal, indexed collection of references and information taken from written and oral orders, messages, journal entries, and meetings. The battle book may include extracts from doctrinal manuals, forms, and other information. It has no specific format and is not a permanent record. (See FM 101-5.) RC UMTs should build battle books in concert with the AC unit they train with in peacetime and/or deploy with for operations.

SUMMARY

3-130. Religious support operations planning, preparation, coordination, and execution must be comprehensive. It must envelop all faith groups and personnel, be synchronized among all commands and UMTs, and be flexible enough to meet the fluid environment. The UMT will be faced with a great deal of information it must manage, implement, and disseminate. Constant and aggressive information gathering and analysis will keep the UMT informed of the religious support requirements for the commander. The chaplain, as a personal staff officer, and the chaplain assistant are a critical assets for ensuring the commander's religious support plan is implemented in a comprehensive and timely manner. Through proper staff procedures, mission rehearsals, and staff coordination, the RSP will meet the mission's requirements.

RESOURCES

3-131. The following is a list of main resources utilized for writing and updating this doctrine.

- AR 165-1.
- JP 1-05.
- JP 5-00.2.
- JP 3-0.

- JP 3-05.
- FM 16-1.
- FM 3-0 (100-5).
- FM 25-100.
- FM 25-101.
- FM 100-17-3.
- FM 101-5.
- FM 55-30.
- National Military Strategy
- Joint Vision 2010
- Army Vision 2010
- Concept for Future Joint Operations, May 1997
- U.S. Army Chaplaincy Strategic Plan FY 2000-FY 2005
- Chaplain Support Force XXI White Paper
- Quadrennial Defense Review Report, September 2001.

Chapter 4

BATTALION, SQUADRON, BRIGADE, REGIMENT AND GROUP RELIGIOUS SUPPORT OPERATIONS

The levels described within this chapter are the smallest echelons to which UMTs are assigned.

CONTENTS

TACTICAL ECHELON DESCRIPTION

BATTALION AND SQUADRONS

4-1. Maneuver battalions can be reinforced with other combat and combat support companies to form task forces for special missions. Task organization increases the capability of maneuver battalions to ensure operational success. A Brigade commander through the staff estimation process may task-organize tank and mechanized infantry battalions by cross-attaching companies between these units. Field artillery (FA) batteries can be reinforced with batteries of any kind to form artillery task forces. Light units and engineer battalions can also be reinforced using this same task organization process.

4-2. Combat support and CSS battalions vary widely in type and organization. They may be separate division or non-divisional battalions but normally perform functional services for a larger unit within that unit's area of operations. Air defense artillery and signal battalions assigned to, or supporting, divisions routinely operate throughout the division area of operations. The Corps Support Battalion (CSB) is part of a Corps Support Command (COSCOM) and a Corps Support Group (CSG) and provides support to units that could belong to a division, or a Corps.

4-3. Battalions, or their equivalent, conduct operations as part of a brigade's operation. Occasionally they are directly under a division or regiment's control. Their capability is increased through task organization. The commander understands the intent of the brigade commander and division commander so that he can properly employ his force.

BRIGADES, REGIMENTS, AND GROUPS

4-4. Brigade size units control two or more battalions or squadrons. Separate brigades of infantry, armor, field artillery, air defense artillery, engineer, and aviation, and armored cavalry regiments can be used to reinforce corps or division and can be shifted from unit to unit to tailor forces for combat.

4-5. Division commanders adjust the organization of their brigades and change their task organization as required by the factors of METT-TC. They normally receive support from FA battalions, engineer battalions (heavy divisions), forward support battalions, and smaller combat, combat support, and CSS units. Their chief tactical responsibility is synchronizing the plans and actions of their subordinate units to accomplish tasks for the division or corps.

4-6. Separate maneuver brigades and armored cavalry regiments have a fixed organization with organic cavalry, engineer, air defense, field artillery, military intelligence, chemical, and CSS units. Separate brigades are capable of operating as independent units.

4-7. Other combat, CS, and CSS brigades and groups are organized to control capabilities for divisions, corps, and larger units. They may also be the building blocks of large unit support structures such as corps and theater army support commands and of combat support commands such as engineer commands. Divisions receive support from an organic brigade-sized, division artillery (DIVARTY), an aviation brigade, and a Division Support Command (DISCOM) comprised of CSS battalions and companies. Heavy divisions have an organic engineer brigade.

Part of a Division

4-8. Brigades or their equivalent usually operate as part of a division. The division is a large, fixed Army organization that trains and fights as a tactical team. Normally, the division commander controls two- to five-ground maneuver brigades. The brigade can be employed in autonomous or semiautonomous operations when properly organized for combat. All brigades must be able to deploy, conduct offensive operations, conduct defensive operations, and conduct retrograde op-

Planning Checklist

Staff: communication, annex, integration, briefings

Specialized:

- Nurture the living:
  - Spiritual Fitness
  - Worship
  - Counseling
  - Visitation
- Care for the Dying:
  - Spiritual triage
  - Leadership
  - Battle Stress
- Honor the Dead:
  - Memorial Services
  - Memorial Ceremonies
  - Rites
  - Sacraments
  - Prayers

erations. Airborne, air assault, and ranger brigades/regiments are capable of conducting forced entry operations. All brigades may deploy to conduct military operations other than war. The brigade UMT is part of the division religious support staff and program.

Part of a Joint Task Force

4-9. Brigades or similar may deploy as part of a JTF with or without its traditional divisional headquarters. In these types of operations, the brigade may work directly for the JTF commander. The demand for experienced liaison will be high for a brigade in a JTF. Liaison may be required with joint, multinational, interagency, and or non-governmental agencies. The brigade's requirement for liaison will exceed its normal personnel and equipment structure.

4-10. Planning Religious Support for this type of operation needs to take into consideration the diversity and complexity of the chain of command, logistics, force structure, and area of operation.

FUNDAMENTALS OF RELIGIOUS SUPPORT OPERATIONS

GOAL OF UNIT MINISTRY TEAMS

4-11. The goal of all Unit Ministry Teams is the personal delivery of comprehensive, timely, and effective religious support to all soldiers.

4-12. Comprehensive, because each soldier has his or her own definition of what constitutes religious support. While not every spiritual need of every soldier can be met in the battlespace, the UMT seeks to meet as many needs as possible. Timely, because time is a limited commodity, and the situation changes in the battlespace. The UMT must be aware of the changing situation and prioritize ministry accordingly. Effective, because religious support must be more than minimal presence and a hurried service by the UMT. Quality effective religious support can provide hope and courage to soldiers who are facing the stress of battle.

ROLE OF THE BATTALION/SQUADRON UNIT MINISTRY TEAM AND SECTION

BATTALION/SQUADRON TACTICAL UMT	
CHAPLAIN (O-3)	**CHAPLAIN ASSISTANT (E-3/4/5)**
Role Commissioned Officer/Staff Religious Professional Title 10 Protected	**Role** Noncommissioned Officer Specialized Assistant Religious Support Specialist Team Security
Mission Advises commander on religious support and quality of life issues to support the unit's mission.	**Mission** Advises Chaplain and NCO Chain. Provides specialized assistance in areas of religious support to support the chaplain.

BATTALION/SQUADRON TACTICAL UMT	
CHAPLAIN (O-3)	**CHAPLAIN ASSISTANT (E-3/4/5)**
Performs/provides/coordinates the religious support mission for the battalion/squadron. Team leader.	Perform battle staff functions. Team member/ NCO.
Parallel Tasks Performs religious support. Provides spiritual, stress, fatigue, suicide preventive counseling to individuals and family members. Plans unit CMRP. Initiates religious support plan. Manages section funds and field offerings. Supervises Chaplain Assistant/Team NCO.	**Parallel Tasks** Prepare and synchronizes religious support and provides emergency religious ministrations in the battlespace. Screens stress, fatigue, suicide individuals for counseling or provide referrals if needed. Implements the unit CMRP. Prepares the Religious Support Annex. Safeguard and accounts for field offerings. Maintains unit equipment.

RESPONSIBILITY AND ACCOUNTABILITY

Staff Responsibilities

4-13. The chaplain serves the commander as a personal staff officer to plan, synchronize, and coordinate religious support, within the commander's area of responsibility. The battalion chaplain is the section leader. The chaplain assistant serves as the section NCOIC.

4-14. Chaplains serve as the "conscience of the command." Chaplains advise the commander on the moral and ethical nature of command policies, programs, actions, and the impact of command policies on soldiers. They advise the commander on—

- Religious and ethical issues as they effect mission accomplishment.
- Indigenous religions and their impact on the mission.
- Accommodation of special religious needs and practices within the command.
- Matters of unit morale as effected by religion.

Relationship to Brigade Unit Ministry Team

4-15. The battalion/squadron UMT works closely with the brigade (BDE) UMT ensuring a comprehensive and coordinated religious support plan for the BDE commander. The BDE UMT technically supervises the subordinate UMT and supports their RSP. During combat operations the battalion (BN) UMTs maintain communications with the BDE UMT sharing information about the religious support mission. In garrison operations, the BN UMT trains with the BDE UMT, coordinates unit and family religious support, and prepares for deployments.

Relationship to the Division Unit Ministry Team

4-16. The battalion/squadron UMT looks to the division UMT as their mentors and advisors during garrison operations. During combat operations, the

BN UMT will be prepared to communicate religious support issues directly if they assume the BDE UMT mission or if they work directly for a division/task force (TF) UMT. The principle of understanding and implementing the guidance of the commander two levels up also applies to the work of the Battalion UMT. The BCT UMT plans its RSP in harmony with the guidance the Division UMT. The Battalion UMT must understand and implement the Division RSP as it plans and executes its RSP.

Supervision Responsibilities

4-17. The chaplain exercises staff supervision and technical control for religious support throughout the battalion AO. The Chaplain is responsible for the development, implementation, and coordination of religious support assets, operations and activities. The chaplain assistant NCOIC is an active member of the NCO support channel in the unit on religious support issues and soldier welfare. The Chaplain assesses the soldiers and other authorized personnel's interpersonal/spiritual well being, which can affect the unit's fighting spirit. The UMT advises the commander on humanitarian aspects of operations, and the impact of command policies on indigenous religions. The Chaplain provides and coordinates privileged and sensitive spiritual counseling and pastoral care to the unit's command, soldiers, authorized civilians, and families. The UMT locates where they can best coordinate, communicate and facilitate religious support across the battlespace.

Staff Supervision During the Operation

4-18. The battalion Chaplain supervises the Chaplain Assistant. The UMT provides direct support and area support to units located within the battalion AO. The UMT coordinates religious support with each unit. The UMT monitors the tactical situation, and plans based on changes in METT-TC. If necessary, they recommend changes to the religious support annex and ensure supported units receive information concerning changes via` FRAGOs. The UMT reports required religious support data to the brigade UMT. The battalion UMT receives operational missions via the Brigade Operational Order/FRAGOs. The brigade UMT is the technical supervisor for the battalion UMT who does prior coordination and establishes contingency plans for the brigade religious support plan.

4-19. Before the battle, the battalion UMT submits a religious support annex that gives guidance to all unit commanders of the religious support provided. The chaplain assistant gathers information that will allow the UMT to battle track the operation, current unit positions, overlays, special operating instruction (SOI) information and successfully navigate the battlespace.

4-20. The UMT before, during, and after a phase of operation will coordinate with the Battalion Task Force for treatment and forward surgery teams to provide comprehensive religious support to soldiers being treated and to the team itself. The UMT coordinates their efforts in conjunction with the movement of the casualties. The UMT maintains contact via radio, casualty feeder cards, etc. with the brigade UMT in the form of SITREPs on their location, health, statistics, and ability to provide religious support such as mass casualty (MASCAL).

4-21. The battalion UMT considers the spiritual condition of the soldiers in follow-on planning. The severity of the previous operation and the number of casualties suffered are key factors in planning, preparing, and executing cycle that is a nonstop process.

Training

4-22. The battalion UMT has the responsibility to be proficiently trained in all areas necessary in order to adequately execute the religious support mission. It is the responsibility of each member of the UMT to be a tactically and professionally proficient team member. They must train in soldier tasks and collective tasks as a team. The battalion UMT trains to the unit's METL utilizing the derived religious support METL. The religious support METL serves as the focal point on which the UMT plans, executes, and assesses training. Training must encompass the following principles to effectively train as a team:

- Train as you provide religious support in the battlespace.
- Use all appropriate doctrine and tactics, techniques, and procedures (TTPs).
- Use performance-oriented training.
- Use mission-oriented training.
- Make training a challenge.
- Train to sustain proficiency.
- Train using multi-echelon techniques.
- Take control and maximize training opportunities.

ROLE OF THE BRIGADE UNIT MINISTRY TEAM

4-23. The religious support section is made up of a minimum of a Unit Ministry Team (UMT) composed of a chaplain (56A) and one enlisted chaplain assistant (56M). The UMT facilitates and coordinates religious support across the battlespace.

4-24. The brigade chaplain is a religious leader, competent staff officer and team leader. The brigade chaplain assistant NCOIC is a leader and advisor, mentor and information manager for the UMT. The UMT must have a keen awareness of the world situation as it bears on potential military operations, must know the role of military religious leaders and must have a broad understanding of the impact of indigenous religions on military operations.

4-25. UMTs must use technology to enhance the personal delivery of religious support. They must be trained in joint and multinational operations and skilled in synchronizing all aspects of religious support during military operations. They will operate in an environment of increasing volatility, complexity and ambiguity, thus the brigade UMT must be innovative and disciplined.

4-26. The brigade chaplain assistant needs to be a graduate of the Battle Staff NCO Course (with the 2S identifier). This will enable the brigade chaplain assistant NCOIC to take part in the MDMP in the absence of the chaplain, to track the battle analytically, and to better utilize the digital enablers that the brigade will share with other staff sections.

4-27. Brigade UMTs require the same level of technology as the command. UMT automation, communication and transportation must keep pace with the command's requirements in order to effectively coordinate religious support, communicate critical information and remain physically flexible to meet requirements across the brigade battlespace. The UMT is an integral part of the brigade.

THE BRIGADE/REGIMENTS/GROUPS TACTICAL UMT	
CHAPLAIN (O-4/5)	**CHAPLAIN ASSISTANT NCO (E-6/7)**
Role Commissioned Officer/Staff Religious Professional Title 10 Protected	**Role** Noncommissioned Officer Specialized Assistant Religious Support Specialist Team Security
Mission Advises commander on religious support and quality of life issues to support the unit's mission. Performs/provides/coordinates the religious support mission for the brigade. Brigade Team Leader.	**Mission** Advises Chaplain NCO Support Channel Provides specialized assistance in areas of religious support to support the chaplain. Performs Battle Staff functions. Team NCOIC.
Parallel Tasks Provides religious support supervision of Bn/Sqd Chaplains. Initiates and develops the religious support plan. Train Chaplains on MOS task. Plan and reviews the BDE CMRP.	**Parallel Tasks** Provide religious support supervision of Bn/Sqd Chaplain Assistant NCOs and Chaplain Assistants. Synchronizes the religious support plan. Trains Chaplain Assistants on MOS task. Prepare and implement the BDE CMRP.

RESPONSIBILITY AND ACCOUNTABILITY

Staff Responsibilities

4-28. The Brigade UMT works directly for the commander. The chaplain serves the commander as a personal staff officer to plan, synchronize, and coordinate religious support, within the commander's area of responsibility. The brigade chaplain is the section leader. The chaplain assistant serves as the section NCOIC. The brigade UMT serves as the technical supervisor and mentor to subordinate UMTs. As the commander's personal staff officer for religious support, chaplains at brigade-level and higher are responsible for assessing the technical proficiency of UMTs in subordinate units. They conduct assessments with the unit during field training exercises (FTXs), Battle Command Training Program (BCTP) events, external Army Training and Evaluation Program (ARTEP) events, Combat Training Center (CTC) rotations, unit training, and command inspections.

4-29. Chaplains serve as the "conscience of the command." Chaplains advise the commander on the moral and ethical nature of command policies, programs and actions, and the impact of command policies on soldiers. They advise the commander on the following, in addition to, the previous battalion/squadron topics:

- UMT personnel issues.
- Division RSP.
- Training issues.

Relationship to the Division Unit Ministry Team

4-30. The Division or TF UMT is the immediate technical supervisor for BDE UMTs assigned or attached operationally to a division or task force. If not assigned or attached, there may be a need operationally for coordination and close communication to ensure comprehensive religious support.

4-31. The BDE UMT or equivalent fully supports the division and/or TF RSP by supervising and providing guidance to BN UMTs RSPs. They work closely supporting the operational information religious support management process for the entire division or task force.

Relationship to CORPS Unit Ministry Team

4-32. The principle of understanding and implementing the guidance of the commander two levels up also applies to the work of the Brigade UMT. It is essential for the BCT UMT to understand and implement the Corps RSP as it plans and executes its RSP.

Staff Supervision During the Operation

4-33. Brigade UMTs provide staff supervision of religious support by visiting subordinate units, monitoring the tactical situation, and planning based on changes in METT-TC. If required, they recommend changes to the religious support annex and ensure subordinate units receive and execute the changes, which are issued as FRAGOs. The brigade UMT should maintain contact with subordinate UMTs by visits, radio spot reports, or any means possible to insure subordinate UMTs status is 100 percent and religious support data is collected and sent to the division UMT.

4-34. The brigade UMT monitors the tactical situation, the status of each battalion UMT, and plans for future operations. If changes in METT-TC warrant adjustments in religious support, the brigade chaplain recommends changes to the brigade commander or executive officer. Reassigning a chaplain or chaplain assistant, or changing the mission of a battalion UMT, requires a FRAGO. The brigade UMT must be prepared to augment TF UMTs in the event of mass casualties. The brigade religious support annex should have a plan to augment the religious support provided at the base support battalion (BSB), forward support medical company (FSMC).

4-35. When the brigade reaches the objective, casualty evacuation increases. The brigade UMT again assesses the requirements for additional religious support in the AO. If the number of casualties is high, a MASCAL is declared. The brigade UMT will issue a FRAGO to notify supporting UMTs. These UMTs will assist the BSB UMT in providing religious support at the FSMC to casualties and by conducting religious support debriefing.

4-36. If a TF UMT has not submitted a current SITREP (See format model in FM 101-5-2), the brigade UMT NCO contacts the team to determine its status. When feasible, the brigade UMT moves forward to the TF UMTs to provide direct spiritual care. Priority goes to the TF executing the main attack. If the brigade continues the mission, time may be very limited. This phase may start with reconnaissance, or it may go directly into deployment for another assault.

4-37. The brigade UMT considers the spiritual condition of the soldiers in follow-on planning. The severity of the previous operation and the number of casualties suffered are key factors in the planning, preparing and executing cycle that is a nonstop process.

4-38. If the brigade is operating as part of a division mission, the brigade UMT of the follow-and-support brigade establishes liaison with the UMT of the lead brigade to facilitate pastoral care casualties and other religious support. The follow-and–support brigade may provide evacuation and treatment for casualties of the lead brigade, so that the lead brigade can continue its tempo. This mission is of particular concern to the brigade UMT because of its complexity, intensity, and momentum.

4-39. In the mobile defense, battalions engaged in offensive, defensive, and retrograde actions present a challenge for the brigade UMT. The UMT plans religious support for each of the units based on METT-TC. The religious support plan is incorporated into the brigade OPLAN as an annex. The area defense differs from the mobile defense in that all units are engaged in defensive operations.

Supervision of Brigade Combat Team (BCT) Religious Support Operations

4-40. All action starts with a vision and that includes religious support. "Vision" is a personal concept of what the organization must be capable of doing by some future point. It is the target. What does the organization need to accomplish today, tomorrow, or years from now to achieve the desired result? How is the organization developed to achieve that capability? Simply put, vision sees the required effort, leadership, and plan to get the job done.

4-41. The transition from battalion to brigade-level religious support requires a transition in focus as well as location. The brigade-level chaplain's focus moves from being primarily an operator to being a staff supervisor of religious support for the commander. A very important part of being a brigade chaplain/supervisor is the development of the vision mentioned above. That vision must broaden to the whole brigade and see supervision as an aspect of religious support proper.

4-42. One component of the vision is certain imperatives for focusing action -- a vision that includes provisions to provide purpose, direction, motivation, and continuity. To provide some insight as to how this applies to the Brigade Unit Ministry Teams, the following few paragraphs will paraphrase FM 22-103, Leadership and Command at Senior Levels.

4-43. Purpose insures there is a clear and worthy object for every action and sacrifice. It provides the UMTs with a reason to withstand the stress of continuous operations. It depends heavily on the ability of the battalion UMTs to understand all requirements. The brigade chaplain does this by giving a sense of priority and importance to necessary tasks. The brigade chaplain explains how the success or failure of the UMTs will impact upon the unit. Teams with a good sense of purpose will succeed in ways that cannot be forecast before the action. Purpose provides the religious support teams of the BCT with goals. It gives the teams a sense of "why" they are doing what they are doing.

4-44. Direction channels the collective efforts of the people on the teams toward the unit's goals, objectives, and priorities. It puts the Unit Ministry Teams on a clear and unambiguous course and thus creates the conditions that allow them to succeed. Setting goals and standards, ensuring discipline, and training the teams are some of the means the brigade chaplain can use to chart a direction for the teams. Goals and standards provide the azimuth. Good discipline provides the confidence that goals and standards are met in the absence of leaders and immediate guidance. Training ensures results. An additional means of providing direction is the SOP. Direction helps the teams know what to do, while purpose helps the teams understand why they do it.

4-45. Motivation engages the will of Unit Ministry Teams to do their utmost to achieve the purposes, goals, objectives, and priorities of the unit, even under the most difficult circumstances of combat. It is not enough to know why and what to do; we need the moral force of a strong will.

4-46. Continuity anticipates future conditions, situations, and needs; it takes steps in the near term to ensure that the BDE Unit Ministry Team can meet future missions. Synchronization of plans and efforts among all Unit Ministry Teams is critical to quality religious support. The Brigade Chaplain is the focal point for this as the staff representative for religious support for the commander.

Supervisory Relationships

4-47. The complex job of the brigade chaplain is to organize the efforts of UMTs in different subunit commands with different battlefield operating system (BOS) responsibilities and support. Some of the teams are in units assigned directly or as attached to the brigade, i.e., the maneuver battalions and field artillery battalion (unit support). To further expand religious support supervision and synchronization, area support to soldiers, members of other services, authorized civilians, and all who have no organic or available religious support is part of the comprehensive plan. The supervisory role expands to denominational coverage to these soldiers. Still, the mandate to UMTs is to provide religious support to all soldiers in the AO.

4-48. The Brigade Chaplain as a staff officer for the commander uses different external coordination channels to ensure the accomplishment of religious support. Three channels of communication command, staff, and technical require development and rehearsal. Too often coordination and guidance are communicated through the wrong channels; and because the Brigade Chaplain has no command authority, he/she must work effectively through the correct coordination channels.

Supervision of Brigade Religious Support Operations

4-49. Task Force UMTs work for their unit commanders. The BDE UMT cannot give guidance to subordinate UMTs that conflicts with their commander's guidance. Proper coordination and solid common sense on the part of the BDE UMT make the entire BDE religious support plan work.

4-50. Proper coordination with the commands and staffs in the brigade opens doors of quality communication and support. A good working relationship builds respect and cooperation with commanders and staffs throughout the brigade combat team. The brigade chaplain and NCOIC need to cultivate this relationship and respect their position on the commander's staff. Proper staffing procedures make religious support a reality.

OBSERVED PRINCIPLES OF EFFECTIVE SUPERVISION
The primary mission of the Brigade UMT is to synchronize Religious Support.
The BDE Unit Ministry Team assigns its highest priority to managing religious support for the whole BDE. The UMT must see supervising as their part of comprehensive religious support. Providing religious support to the tactical operations center (TOC) is secondary to the primary mission of synchronizing religious support throughout the BDE. When the BDE UMT is effectively overseeing religious support, quality religious support reaches all corners of the BDE.
PRINCIPLE 1. The BDE UMT trains as a team before it arrives in a theater of operation. A goal of the BDE UMT training is to work as a synchronized team. UMTs do individual religious support well, but the BDE UMT must work hard at coordinating the religious support that happens in the Brigade.
PRINCIPLE 2. Synchronize Religious Support Appendices at all levels. The BDE UMT ensures that religious support is synchronized on paper and in action. Make certain that the Brigade Religious Support Appendix is accurately reflected in subordinate UMT Religious Support Appendices. A goal for BDE UMT training is to have continuity in religious support appendices throughout the BDE UMT.
PRINCIPLE 3. Plan, rehearse and coordinate casualty replacement plans for members of the BDE UMT. UMT members in the BDE should know who would replace them in the event they become a casualty.
PRINCIPLE 4. Coordinate all Religious Support guidance with the appropriate commanders. Just because something is written in an appendix does not mean that it will happen, it must be coordinated. Coordinate religious support prior to the mission with all subunit UMTs. This effort helps to ensure comprehensive religious support synchronization in the AO.

OBSERVED PRINCIPLES OF EFFECTIVE SUPERVISION
PRINCIPLE 5. Daily communication is vital to Religious Support. UMTs have a crucial piece of the overall religious support plan and need to send that information up and down. UMTs must know what sort of information is necessary to report. BDE UMTs must be aggressive in maintaining disciplined and systematic information flow. Daily communication between all UMTs in the BDE is imperative for up-to-date information. Here are some suggestions for making that work: • Net calls. A net call establishes a specified frequency and time when all UMTs are on the radio at the same time. In this way information can be put out quickly and every UMT becomes knowledgeable of the religious support situation. • Maintain regular hours for monitoring the radio and mobile subscriber equipment (MSE) so subordinate UMTs know when they can directly contact the BDE UMT. • Send daily guidance and updates to subordinate UMTs. • Spot reports to pass on time sensitive information. • Use the BDE RETRANS frequency (coordinate with the signal officer. • Send messages by couriers such as commander's drivers, ambulance drivers, and logistics package (LOGPAC) drivers. • Authorize lateral communication between UMTs. Often one UMT cannot make radio-communication with the BDE UMT, but can pass a message through a neighbor UMT. • Schedule regular meetings with subordinates in the field.
PRINCIPLE 6. Daily updates from the BCT UMT to subordinate UMTs are critical. When the BDE UMT communicates to its subordinate UMTs it builds teamwork. This enables individual UMTs to understand what is happening, and what is expected of them. They feel like part of the team. Always remember that effective communication is a two way street. UMTs must communicate the following items of information: • A daily status report of all UMTs in the BDE. • UMT NCOs update. • Changes in religious support for attached/detached units, and units operating in the brigade AO. • Upcoming religious events in the command, and indigenous religious holidays. • The Brigade Commander's daily guidance and other important issues from the command. • Questions that need answering reference indigenous religions and their potential impact on operations.
PRINCIPLE 7. Daily SITREPs from subunit UMTs to the BDE UMT are critical. The BDE UMT must know what the subunit UMTs know to effectively manage the religious support mission. Emphasis upon effective communication will build teamwork and enhance the religious support mission. Examples of information that must be communicated follows: • The status of the UMT. • The status of ecclesiastical supply. • Daily activity. • Changes in unit RPP. • CA issues. • Information on indigenous religious leaders and sacred sites. • Issues of religion, morals, ethics, and morale.

TRAINING

4-51. To supervise religious support in a combat environment, the BDE UMT creates well-rehearsed coordination, synchronization, and communication techniques through training before deployment. If it does not happen effectively in a garrison/training environment then, quality religious support may not happen in a stressful and fast-paced combat operation. The responsibility for training rests heavily on the BDE UMT. Quality combat survivability training is available at the unit level. Religious support training is available at division and higher level. This does not prevent BDE Unit Ministry Teams from conducting training within the BDE. Training at this level is crucial to support the commander's mission. The BDE UMT needs to be proactive with an approved, staffed and coordinated training plan that takes into account Officer and Enlisted Critical Task lists as part of the total BDE training plan.

4-52. In addition to the BN training topics and individual and collective responsibilities, the BDE UMT must be personally involved in the training of subordinate UMTs. The BDE UMT plays a critical role in the development of the subordinate unit's junior chaplains and chaplain assistants. Success in providing spiritual support to soldiers requires adequate training. The UMTs of a BDE require sufficient training to enable it to mobilize, deploy, and provide quality spiritual support to soldiers anywhere in the world. The battlespace fixes the directions and goals of training. The BDE UMT is the training management team for subordinate UMTs. Their responsibility is to initiate, plan, and coordinate training for all UMTs in the command. The BDE UMT will make training effective by:

- Developing and communicating a clear training vision.
- Requiring subordinate UMTs to understand and perform their roles in unit, technical, and individual training.
- Being proficient on their mission essential tasks.
- Continually assessing training completed.
- Eliminating training distractions.
- Developing near and far term training plans.
- Being systematic and aggressive.

4-53. Suggested training topics include—

- Land navigation—mounted/night/navigation.
- Operations order process—
  - Developing and writing a religious support estimate.
  - Writing a religious support annex.
  - Writing religious support FRAGOS.
- War-gaming a battle as a brigade UMT combat team.
- Communications rehearsal—
  - Daily reports.
  - Communicating vital information.
  - UMT net call.
  - Practicing alternate means of communication.
  - Practicing cross communication between BN UMTs.
- Rehearsing the RSP of succession.

- Re-supply training: ordering and getting ecclesiastical supplies through logistics channels.
- Map reading, military symbology, and battle tracking.
- Conducting Religious Area and Impact Assessments
- UMT TSOP development and rehearsal.
- Chaplain assistant coordination of the RSP in absence of the Chaplain.
- Briefing the RSP.
- Staff Integration.

OPERATIONS PLANNING AND PREPARATION FOR UNIT MINISTRY TEAMS

4-54. Military operations consist of four types: offensive, defensive, stability and support. Each type consists of four phases: pre-deployment, deployment, sustainment, and redeployment. The following is a discussion of the responsibilities of company and field grade UMTs in each type and through each phase of operation.

CREATING THE TASK FORCE RELIGIOUS SUPPORT PLAN

TIMELINE OF CRITICAL EVENTS + PRIORITY OF RELIGIOUS SUPPORT = TF UMTs RSP

Timeline

4-55. The most valuable resource is time. The UMT must manage time so that the RS mission gets accomplished. A TF chaplain will not be able to see every soldier and certainly not solve every problem. Set priority of effort and allocate time to that effort. You must use the tools of time management. First, the UMT extracts all the time critical events from the OPORD and places them on a time line; for example—

- Logistics package (LOGPAC) activation and deactivation.
- Enemy recon expected.
- Friendly recon/counter-recon sent out.
- TF rehearsals/brief-backs.
- CSS rehearsals.
- Engineer movements.
- Company team movements.
- Line of departure (LD) /Be-Prepared-to-Defend times.
- Jump Aid Station (JAS) moves out, and any other time driven events.

4-56. Second, the UMT determines the intent of religious support during this phase of the battle. Knowing the religious support intent will help determine how much time to spend on visitation, where to spend it, and where priorities lay.

4-57. Third, the UMT determines soldier activity level. The Commander's Intent and unit warfighting doctrine must be understood. With these in mind, the UMT will know what tasks the company teams must accomplish. The UMT may then gauge when is the most effective time to visit and provide religious support.

4-58. Fourth, the UMT determines limiting factors like restrictions in the OPORD, ROE, METT-TC, and personal state of readiness.

4-59. The RSP is an informal plan and is flexible to meet mission requirements. It is likely to change during the execution phases of the battle because not everything goes as planned. It is easier to alter an existing plan than create (and communicate) one in a crisis.

Priority

4-60. Priority of religious support is to be identified as early as possible.

4-61. First, the UMT takes into account what has been experienced by each company in the TF, and their role in the coming mission. The UMT must determine what and where the needs are, and which unit has priority for religious support.

4-62. Second, the UMT considers morale factors known before the battle, leadership issues, human sustainment issues (food/rest/water), and interpersonal conflicts.

DEVELOPING PRIORITIES FOR THE TASK FORCE UNIT MINISTRY TEAM

4-63. The driving force behind the planning process of a TF UMT is determining its religious support priorities. Where will you place your main efforts and why? The UMT must create a time line that includes the unit's critical-time-driven-events. Working with this data, the UMT can then select blocks of time for priority religious support.

1. ***Study the Brigade OPORD/OPLAN and extract the following information:***
   - The mission assigned to your battalion.
   - The BDE Commander's intent for this mission.
   - Attachments / Detachments
   - The RS missions assigned to the UMT - Unit and Area Support.
   - Channels of communication with the BDE UMT, and with UMTs in the BDE area of responsibility (AOR).
   - Succession plans in case the BDE UMT becomes a casualty.

2. ***Study the Battalion OPORD / OPLAN and extract the following information:***
   - The missions assigned to the various Battlespace Operating Systems (BOS).

– Intelligence	(S2 / G2/Scouts)
– Maneuver	(combat companies)
– Fire Support	(artillery, mortars, close air support)
– Air Defense	(Air Defense Artillery)
– Mobility & Survivability	(Engineers)
– Logistics	(support, supply, repair, maintenance; medical)
– Battle Command	(Command and Staff)

3. Identify the main effort company(s), the supporting attack company(s), and the Scout's mission.	
4. Identify the first location for the medical units (BAS, JAS, FSMC), and the next several places each will jump by phase. Locate the AXPs and plot them on your map.	
5. Determine the priority of religious support the UMT will perform:	
• Main effort company(s)	
• Supporting attack and reserve company(s)	
• Ordinances/ Rites / Sacraments / Worship services	
• Visitation / Counseling	
• Senior leader religious support	
• Company / Platoon leader religious support	
• Individual soldiers	
• Morale problem areas	
• Honor the dead - Memorial Ceremony/Service	
• Scouts	
• Area Support requirements	
• Religious support to other UMTs	
• Ethical / Moral counseling	
• Displaced Persons / Detained Person religious support	
6. Create a time line for your religious support	
• Write in the critical events of the day that effect the UMT.	
• Identify blocks of time for priority religious support.	
• Write in your priority religious support objectives in the priority religious support time blocks.	
• Consider flexible time factors in a high operation tempo (OPTEMPO) and fluid combat environment.	
• Choose an alternative plan in case conditions change unexpectedly.	

WRITING RELIGIOUS SUPPORT APPENDIX

4-64. ***Consider the audience.*** The religious support appendix to the Service Support Annex is for the entire command. The religious support appendix must contain information that is both specific to subunit UMTs, and important to the whole command. The religious support appendix must support the commander's mission.

4-65. ***Only submit the appendix when it's needed.*** There are many times when there is no critical information to contribute. When appropriate, submit an appendix/matrix or a paragraph for the CSS annex. (See Appendix D)

4-66. ***Content is key, not the format.*** The UMT has no command authority so all tasks must be coordinated with commanders. Working with the S3 / G3, insert tasks for subordinate UMTs into the main body of the OPORD as tasks to subordinate units in paragraph 3. This will ensure command awareness and support for these tasks. Space for the religious support appendix varies from three pages in the BDE OPORD; one page for a matrix; or possibly only have space for a paragraph. Use the correct military format. A matrix can be utilized in keeping with the styles and formats used in the unit (See Appendix D). Or write a concise paragraph for inclusion in the CSS Annex. Ensure the content only states what is critical to support the mission.

4-67. ***A good TSOP reduces OPORD size.*** TSOP material is not repeated in the RS Appendix! The TSOP is an effective, usable, and combat oriented document. The content of the TSOP will help in the preparation of appendices and pre-combat coordination. UMTs must possess the RS portions of higher and lower units TSOPs.

4-68. ***Intent should be clearly written.*** Write the concept of religious support so higher and subordinate UMTs understand what is to be accomplished.

BRIGADE

4-69. The primary mission of the brigade UMT is to coordinate religious support for all soldiers and authorized civilians in the brigade commander's area of operations (AO). They provide religious support to soldiers in the brigade headquarters and others in the AO without religious support assets. They provide critical guidance and technical control to subordinate UMTs.

RELIGIOUS SUPPORT COVERAGE PLAN

4-70. The brigade UMT coordinates the brigade UMTs religious support throughout the brigade AO, thus ensuring that all units and individuals operating within the AO are provided religious support. This includes refugees, displaced persons, civilians in the area of operation, and enemy prisoners of war. The support of civilians and Enemy Prisoners of War is the responsibility of the commander; the commander may or may not require the chaplain to provide or perform religious support. As part of the Army component, civilian employees and contractors will be present providing essential work. Pending higher guidance issued in the division OPORD, the UMT will assess the requirements of civilian employees and contractors deployed in the AO and will include them in the religious support plan, at all levels.

CONTRACTORS

4-71. Chaplains are the staff officers in charge of coordinating religious support for contractors. All contractors who are clergy/lay ministers and readers will submit an application for approval as prescribed under Army Regulation (AR) 165-1 for supervision and coordination by military chaplains. UMTs will make every effort to plan for adequate religious support resources with their unit such as literature and sacramental supplies to support the religious needs of contractors. UMTs will plan for area religious support coverage to include contractors. Coordination for additional ecclesiastical supplies or religious support personnel will be with the supervising UMT. Chaplains will provide support for death notifications, Red Cross notifications by command, liaison with CONUS clergy, or others for contractor casualties. Religious support is coordinated through the chain of command.

RELIGIOUS SUPPORT MISSION TASK

4-72. Before the battle, the supervisory UMT submits a religious support appendix that both gives clear guidance to subordinate UMTs and allows all unit commanders to be informed of the religious support provided. The chaplain assistant gathers information that will allow the UMT to battle track the operation and current unit positions, understand overlays and SOI information, and successfully navigate the battlespace

4-73. The brigade UMT also coordinates with combat support and combat service support units in the brigade AO. Units with organic UMTs may need additional faith group support. Some UMTs may provide religious support for units whose organic support is not available. Time, distance, and battlespace conditions may restrict the travel of the UMTs and make religious support difficult.

Preparation for Combat

4-74. The brigade UMT uses the religious support estimate to plan religious support for all units in the brigade AO. It assesses the faith group needs of the battalions and requests additional support from the division UMT. The brigade UMT coordinates times and locations for forward link up points with the maneuver battalions.

4-75. The brigade UMT coordinates also with combat support and combat service support units in the brigade AO. Units with organic UMTs may need additional faith group support. Some UMTs may provide religious support for units without organic religious support or for units whose organic support is not available. Time, distance, and battlespace conditions may restrict the travel of the UMTs and make religious support difficult.

4-76. During offensive operations, the brigade support area (BSA) moves frequently. The brigade UMT may establish an alternate operational base in the BSA, but it must know how to move its equipment if the BSA relocates. As an alternative, the team may choose to establish its operational base near the brigade tactical operations center (TOC), where the majority of planning and communications are found. In the final analysis, the BDE UMT must locate in a position that best allows it to facilitate RS across the brigade. Regardless of location, the brigade UMT must be prepared to displace quickly.

The brigade UMT attends rehearsals to ensure religious support is fully integrated with the scheme of maneuver and the CSS plan.

4-77. During rehearsals, the brigade commander ensures the BSB commander has chosen adequate main supply routes (MSRs), future locations, logistics release points (LRPs), and refueling sites. The brigade UMT monitors this part of the rehearsal closely to integrate religious support with the scheme of maneuver and CSS plan.

4-78. The brigade UMT monitors the tactical situation, the status of each battalion UMT, and plans for future operations.

Develop Intelligence through reconnaissance and intelligence assets thru Exercise Command and Control

4-79. The brigade UMT monitors the tactical situation, the status of each battalion UMT, and plans for future operations. If changes in METT-TC warrant adjustments in religious support, the brigade chaplain recommends changes to the brigade commander or executive officer. Reassigning a chaplain or chaplain assistant, or changing the mission of a battalion UMT, requires a FRAGO. The brigade UMT must be prepared to augment TF UMTs in the event of mass casualties.

Perform Logistics and Combat Service Support

4-80. When the brigade reaches the objective, casualty evacuation increases. The brigade UMT again assesses the requirements for additional religious support in the AO. If the number of casualties is high, the team may choose to move to the forward support medical company in the BSA.

If the number of casualties is high, the brigade UMT may choose to move to the forward support medical company in the BSA.

4-81. If a TF UMT has not submitted a current SITREP, the brigade UMT NCOIC contacts the team to determine its status. When feasible, the brigade UMT moves forward to the TF UMTs to provide pastoral care. Priority goes to the TF executing the main attack.

4-82. If the brigade continues the mission, time may be very limited. This phase may start with reconnaissance, or it may go directly into deployment for another assault.

4-83. The brigade UMT considers the spiritual condition of the soldiers in follow-on planning. The severity of the previous operation and the number of casualties suffered are key factors in planning.

4-84. Split Operations could include a combined Legal, Medical, UMT and Civil Affairs Cell as an active battle-tracking proponent in TOC operations, while a separate religious support cell, employing all BSA UMTs, operates in the BSA TOC. This type of operation promotes better use of RS assets, prepares the BSA Chaplain to assume the responsibility of the TF Chaplain in the event of loss, provides greater opportunities to minister to TF BOS systems, and produces two avenues for receiving TF UMT daily reports. The TF UMT maintains contact with the BSA cell through digital nonsecure voice telephone (DNVT) channels.

4-85. If the brigade is operating as part of a division mission, the brigade UMT of the follow-and-support brigade establishes liaison with the UMT of the lead brigade to facilitate pastoral care to casualties and other religious support. The follow-and-support brigade may provide evacuation and treatment for casualties of the lead brigade, so that the lead brigade can continue its tempo. This mission is of particular concern to the brigade UMT because of its complexity, intensity, and momentum.

Division Cavalry Squadron

4-86. The division cavalry squadron may operate across the entire division front or along one or both flanks. When the squadron operates forward of the brigades, the squadron UMT coordinates with the brigade UMTs for the care of casualties and for additional religious support.

DEPLOYMENT

4-87. BDE UMTs are prepared at all times to deploy with their respective unit. Their deployment plan is synchronized and coordinated with the Brigade command and staff, subordinate TF UMTs, and higher supervisory UMTs. The UMT trains throughout the year on the needed skills and tasks for deployment. Planning considerations include—

- Air/Ground flow schedule.
- Intelligence information.
- Subordinate UMT readiness.
- Unit spiritual fitness.
- Ecclesiastical resupply kits and other religious support material.

Pre-combat Planning Considerations

4-88. After completing the planning process, the UMT is not yet ready to begin religious support operations. Coordination with key people and pre-combat checks still need to be completed. Listed are examples of areas for coordination, inspection, and team member responsibility.

- Chaplain—
  - Worship/Counseling preparation.
  - Coordination with—
    - Executive Officer for religious support plan/appendix; staff meetings; daily activities.
    - Commander for his guidance and focus; give advice on spiritual fitness of command.
- Chaplain Assistant—
  - Coordination with headquarters and headquarters company (HHC)/headquarters and headquarters troop (HHT) commander for bivouac site; local security plans; food and fuel; planned moves of trains or TOC.
  - Checks and Activities—
    - Individual precombat inspection (PCI).
    - Weapon/ammo.
    - Mission oriented protective posture (MOPP) gear.
    - Survival gear.
    - Food.
    - Water.
    - Issue equipment.
  - Vehicle PCI—
    - Meals ready to eat (MREs) stored.
    - Operator's vehicle material (OVM) stored.
    - Fuel.
    - Water.
    - PMCS complete.
    - Radio and frequencies, crypto, fill.
  - Team PCI—
    - Map with overlays.
    - CEOI/SOI information.
    - Night Vision Goggles with spare batteries.
    - TSOP.
    - Compass.

        - M256 kit.
    - Movement Plan—
        - Map Recon.
        - Plan posted in Combat Trains Command Post (CTCP)/Field Trains Command Post (FTCP)/Tactical Operations Center (TOC).
    - Ecclesiastical Supplies—
        - 30/60/90 day supply.
        - Chaplain's Kit complete.
- Unit Ministry Team—
    - With subordinate unit commanders/CSMs/1SGs—
        - Plan subordinate unit worship services.
        - Identify times and locations for visiting soldiers.
        - Identify special spiritual needs of the command.
        - Identify other special needs and issues.
    - Higher and lower-level ministry teams—
        - Religious Support Plan (at least two levels up).
        - Location during phases of battle.
        - Religious support requirements.
        - Availability of and plans for denominational support.
        - Emergency ecclesiastical re-supply.
        - UMT Replacements.
    - Staff—
        - Casualty information.
        - Intelligence updates (critical for the UMT to keep themselves constantly updated).
        - METT-TC updates.
        - Medical Plan updates.

OPERATIONS

Staff Coordination

4-89. Before departing the operational base, the UMT leaves a copy of its travel plan in the command post (CP) or TOC. The chaplain coordinates with the S1/S4 representative to receive messages and requests for religious support. The brigade chaplain is normally expected to attend shift change and battle update briefings (BUB). The brigade chaplain should ensure a means whereby those who try to contact the UMT can leave messages and that the UMT location is known at all times.

4-90. During offensive operations, the brigade support area (BSA) moves frequently. The brigade UMT may establish an operational base at the brigade TOC. The BDE TOC provides the UMT—

- Current operational updates.
- Availability to the commander and key staff personnel.
- Battle tracking.

4-91. As an alternative, the team may choose to establish its operational base in the BSA. The TOC is preferable. Regardless of location, the brigade UMT must be prepared to displace quickly.

4-92. During rehearsals, the brigade commander ensures that the Brigade Support Battalion (BSB) commander has chosen adequate MSRs, future locations, re-supply points, and refueling sites. The brigade UMT monitors this part of the rehearsal closely to integrate religious support within the scheme of maneuver and CSS plan.

Note: When moving independently among task force and slice units, the brigade UMT must be aware of the potential for being left behind. To prevent this situation, UMTs must have dedicated communications equipment that is compatible with the unit's secure communication equipment.

Coordination with the CSS planners
1. Refer to FM 71-2, CSS chapter.
2. The BDE will issue its OPORD to the TF leadership early. The BDE OPORD contains the BDE UMT's Religious Support Appendix. Coordinate with your S4, and ensure his familiarity with the religious support appendix.
3. The TF Cdr will issue a warning order to the TF. A possible format is in FM 71-2, appendix
4. At this point one of two things might happen; either the S4 will return to the CTCP and convene a meeting with the battalion maintenance officer (BMO), medical platoon leader, S1, and the Chaplain to review the BDE order and receive feedback for CSS planning; or the S4 stays at the TOC and holds the CSS planning meeting there. If the Chaplain is occupied with ministry, the chaplain assistant attends the CSS planning meeting and provides UMT input.
5. The Cdr will issue the OPORD to the command at the TOC or other designated location. One member of the UMT should attend that meeting.
6. After the UMT has studied the OPORD, coordinate with the medical platoon leader. Find out first hand what the CSS plan is, especially the planned locations during each phase of the battle.
7. Attend the CSS rehearsal and listen for changes in the plan. During the rehearsal, let everyone know where the UMT will be during each phase of the battle.
8. Monitor the Admin/Log net to stay current. Often the Command puts out changes to the plan on short notice. The tactical situation changes rapidly and you might find the planned attack has changed into a hasty defense. Be alert, your life depends on it.

SITUATIONAL AWARENESS

4-93. Frequently the operation does not take place as planned. The UMT must know how to get information from a number of sources in the unit. There will be times when the back-channel information flow is more timely and accurate than the official flow.

4-94. Timely coordination with the CTCP leadership (S4 and S1) and their staffs before deployment makes it easier to coordinate and publish UMT FRAGOs. If the CTCP has not planned to track the UMT before it goes to the field, then in all likelihood it will not track the UMTs movements and priorities when in the field. Prior coordination makes it happen.

4-95. If the UMT announces that it will visit a particular company or section, then it needs to follow through on that plan. The UMT must carefully consider any information it places in the TF OPORD.

4-96. The Chaplain is a key player on the CSS staff. There are critical missions the chaplain must accomplish. The UMT must integrate with the CSS staff before deployment and understand all the forms, reports, and planning process used by the staff. The TF XO is responsible for all CSS within the TF and the chaplain must insure the XO is familiar with the critical religious support coordination factors that must be completed before deployment.

4-97. Finally, the CSS staff will produce estimates of anticipated casualties and expenditures for the upcoming battle. These estimates help the UMT to plan their religious support mission more effectively.

THE BRIGADE SUPPORT BATTALION UNIT MINISTRY TEAM IN THE BRIGADE SUPPORT AREA

4-98. The Brigade Support Area (BSA) is generally located behind the rear boundary of the TF sector. The BSA contains the support elements for the entire Brigade. The Engineer Sapper Battalion, the Brigade Support Battalion (BSB), and the field trains for all the task forces are just a few of the CSS assets in the BSA (FM 71-3). Normally, the TF UMT rarely visits this area because of the distance from the TF fight. There will be a supply route (SR) from the BSA into each TF sector. Each BDE and TF Chaplain should read the CSS chapter in FM 71-2 to gain an understanding of the composition and doctrine of the Field Trains in a BDE.

4-99. The HHC Commander is in charge of the Field Trains and will have a Field Trains Command Post (FTCP). The FTCP has the job of tracking the battle and coordinating the CSS actions in the BSA. The support platoon locates here (including mess and maintenance) and thus the LOGPAC system for the TF. The HHC Commander will also attend the "tenant's meetings" given by the BSB leadership to coordinate CSS issues.

4-100. Communication with the FTCP is difficult due to distance. The best method is to determine how the S4 communicates with the FTCP and use the same method. Possible pathways of communication are: the mobile subscriber equipment (MSE); frequency modulation (FM) radio communications with the TF A/L (Administrative / Logistics) net are possible if the distance is not too great. Sometimes a messenger can be used. Individuals to carry messengers can be found on the LOGPAC convoy. Link up with the LOGPAC convoy at the Logistics Release Point (RLP) in the TF and meet with the Support Platoon Leader or Platoon Sergeant.

4-101. The FTCP Command and Staff must understand that the BDE UMT or another UMT might work with them on various issues. For example, the BSB UMT might have the mission of working with the replacements going forward before their assignment to the individual companies. The BDE UMT might use the FTCP to pass messages/reports forward to the TF UMT and vice versa. One of the UMTs in the BSA normally provides religious support for the soldiers in the field trains.

4-102. The BSB is a critical asset for the BDE. The medical company will provide evacuation assets to the TF medical platoon as well as serve as a replacement treatment team if needed. The flow of ambulances between the BSB medical company and the TF BAS provides a readymade transportation/communications network. Companies evacuate wounded to a Patient Collection Point where an ambulance from the BAS/JAS collect the wounded. The wounded receive emergency treatment at the BAS/ JAS. A BSB medical ambulance will then evacuate them directly or through an ambulance exchange point (AXP). The soldier receives further treatment in the BSB before further evacuation. The BDE chaplain will assign a UMT to work with the wounded in the BSB (normally the BSB UMT). Communications can go by the evacuation route. If the TF UMT uses this method, make prior coordination so BSB ambulance personnel understand their role in passing messages, and the BDE UMT arranges to get the messages. This should be in the TSOP for both the TF and the BSB to insure no misunderstanding. Again, the TSOP is very important!

4-103. Tenant's meetings held at the BSA by the BSB leadership are for all elements of the BSA community. A representative from the BDE S1/S4 command post will normally attend as well as the TF HHC Commander from the TF Field Trains Command Post. The BDE UMT normally has its base of operation in the BDE S1/S4 command post area. Information and messages may go with the TF HHC Commander to the BDE S1/S4 representative.

THE ENGINEER BATTALION

4-104. The Engineer UMT is usually based in the BSA. Engineer activities are directed through the engineer Administrative/Logistics Operations Center (ALOC) which is located near the BDE TOC. The Engineer UMT should assume some of the area RS requirements in the BSA. Normally, the engineer assets in the TF are in known locations at a Class IV point. If the Engineer UMT elects to move forward into the TF sector to visit its scattered unit, it should first coordinate with the BDE UMT.

4-105. Religious support within the BSA and for the entire BDE is critical. The BSB UMT should have a definite RSP for their area of operations and for the BDE through the BDE UMT. The BDE UMT should provide specific guidance and tasks for the BSB UMT. Some tasks are as follows—

- Provide religious support to the medical staff.
- Read casualty feeder cards to determine if religious support has been given by forward TF UMTs.
- Receive/send messages from/to TF UMTs then relay messages to BDE UMT.
- Brief and prepare replacements for combat operations.
- Provide BDE UMT information on soldier replacement issues, medical trends, number and types of casualties, and the BSA situation.

RELIGIOUS SUPPORT TO CASUALTIES

4-106. Spiritual care to wounded soldiers is one of the core functions of the UMT. It contributes to the soldiers' well-being and recovery. Prayers, scripture readings, rites, ceremonies, and sacraments constitute religious support for wounded or dying soldiers. While the chaplain gives care to individual casualties, the chaplain assistant moves among the other soldiers to provide care and to identify those in greatest need of assistance.

4-107. During combat operations, maneuver battalion UMTs may locate with the battalion aid station. If the medical platoon divides the aid station into two or more treatment teams, the UMT should consider locating where the most casualties are expected. Based upon experience, the UMT may split for a short time to maintain contact with the separate treatment teams. However, this is based upon training, experience, and necessity. The chaplain must carefully weight all aspects.

4-108. The UMT considers the physical condition of the casualties and gives priority to the most seriously injured. The following medical categories help the team determine priorities for care:

- ***Expectant:*** casualties with very serious injuries and low chance of survival. The UMT provides pastoral and spiritual care for the dying as its first priority.
- ***Immediate:*** casualties with a high chance of survival if medical measures are accomplished quickly.
- ***Delayed:*** casualties needing major surgery, but who are able to tolerate a delay in surgical treatment.
- ***Minimal:*** casualties with relatively minor injuries that can be treated with self-care. Do not ignore these casualties as most will be treated and returned to combat with potentially high levels of anxiety.

RELIGIOUS SUPPORT TO CAREGIVERS

4-109. Religious support to caregivers is a vital part of the UMT's mission. Often the gravity of the situation, the need to give aid quickly, and the absorbing engagement of the moment will carry caregivers through the crisis. When the crisis is concluded, however, the impact of what has happened may strike the caregivers suddenly and dramatically. Medical personnel and mortuary affairs personnel may be in particular need of religious support at the conclusion of a traumatic event.

RELIGIOUS SUPPORT TO GRIEVING UNIT MEMBERS

4-110. Religious support to everyone involved in a military operation is critical. Often survivors are stunned and immobilized by the loss of fellow soldiers. The emotional and spiritual impact of loss requires the finest skills of the UMT to deal with the denial, anger, guilt, and fear, and to facilitate the continuation of life and service.

4-111. Care should begin as soon as possible after the loss occurs. The knowledge that soldiers have received religious support gives comfort and encourages grieving soldiers to move on with their lives. Following the immediate intervention, supportive follow-on care should be provided when soldiers re-deploy to their home station.

OFFENSE

4-112. The offense is the decisive form of combat operations. The main purpose of the offense is to defeat, destroy, or neutralize the enemy force. While strategic, operational, or tactical considerations may require defending for a time, defeating the enemy requires shifting to the offense. Rapid movement characterizes offensive operations over large areas. The types of offensive operations are movement to contact, attack, exploitation, and pursuit.

4-113. Offensive operations place great demands on UMTs. The constant movement of units and troops, and the operation tempo (OPTEMPO) of the battle limit the UMTs' access to soldiers. The religious support focus will vary from mission to mission. Once an offensive operation begins, UMTs in combat units will provide ministry on a different scale and at different times from those in combat support and combat service support units.

OFFENSIVE PLANNING CONSIDERATIONS

4-114. The UMT plans religious support for each phase of an operation. Offense planning considerations include—

- Develop Intelligence through reconnaissance and intelligence assets.
- Conduct Maneuver.
- Deployment.
- Employ Firepower.
- Protect the Force
- Exercise Command and Control (C2)
- Perform Logistics and CSS Consolidation and Reorganization or Continuation.

(See FM 100-40 for more detail.)

INFANTRY OR ARMOR BATTALION/TASK FORCE

Preparation for Combat

4-115. The Task Force (TF) UMT begins planning for religious support by considering METT-TC. The mission and the time available will determine how the UMT provides religious support. With little time, the team plans and prepares quickly not to miss opportunities for religious support to soldiers. For example, the time available for religious support before a hasty attack is much less than the time available for religious support before a movement to contact or a deliberate attack.

4-116. The UMT carefully establishes religious support priorities. It considers the order of march, the units conducting the main and supporting attacks, and the severity and number of casualties in previous missions. Priority is given to elements that have sustained the most casualties or to those that will be engaged at the earliest. Pastoral care to casualties is always the priority. Therefore, the UMT coordinates with the medical platoon when casualty evacuation plans are formulated.

Develop Intelligence through Reconnaissance and Intelligence assets.

4-117. This phase begins as soon as possible after the unit receives a mission. Reconnaissance elements seek to detect enemy activity and the physical characteristics of areas of interest to the commander.

The scout platoon is an early priority for religious support.

4-118. The TF scout platoon begins its reconnaissance and surveillance as soon as the mission is received from the brigade. It may leave as much as 24 hours before the main body crosses the LD. The scout platoon is an early priority for religious support.

4-119. While the scout platoon conducts the reconnaissance, the rest of the unit continues preparation for combat. This preparation normally takes place in a tactical assembly area (TAA). Worship services, sacraments and ordinances, individual counseling, and visitation are possible and very important for soldiers during this phase. The UMT coordinates with company commanders for times and places for religious activities. An excellent opportunity for this coordination is the meeting at the LRP.

> The UMT coordinates with company commanders and first sergeants for time and place of religious activities.

4-120. The UMT utilizes the intelligence preparation of the battlefield (IPB) process by the S-2 staff section to identify any aspect within the AO that will affect the manner in which they accomplish the religious support mission.

Conduct Maneuver

4-121. When attacking from positions not in contact with the enemy, units normally stage in a TAA. The TF does not move all at once. While the line companies move toward the LD, the main command post, the combat trains, battalion aid station (BAS), support elements of the TOC, mortars, and other elements of the TF prepare for the operation. Religious support is given to each element of the TF before it begins movement. Soldiers who are preparing for the operation may not have time to gather for worship.

> The Task Force UMT normally travels with the combat trains or with the BAS.

4-122. The TF UMT normally travels with the combat trains or with the BAS. If the team is visiting soldiers in other locations before movement, it must determine ahead of time which routes will be clear of obstacles, how long it will take to return to the BAS, and the location of link up points. The UMT times its movement to precede or coincide with the scheduled start time of the BAS. All movement is coordinated with the S4.

4-123. If the medical platoon intends to split the BAS, the team must plan support for more than one site. The UMT must know where the TF surgeon will locate; where most of the casualties are expected; and, if the UMT needs to separate, how it will maintain two operations.

4-124. The displacement of critical CSS elements, such as the combat trains, BAS, Unit Maintenance Collection Point (UMCP), Logistic Release Points (LRPs), and Casualty Collection Points (CCPs) will be triggered by the TF crossing phase lines or by other control measures. This means of movement allows the CSS elements to remain responsive to the task force. The TF UMT monitors the tactical situation to know when the TF is approaching each phase line and thus when it may need to move or shift its religious support focus.

4-125. Because of distances and the OPTEMPO, most units will re-supply on the move. The TF UMT will plan accordingly for its own re-supply.

4-126. As the TF moves, communication becomes difficult. The UMT will find it hard to monitor the tactical situation. Good planning, rehearsals, and the ability to adapt on the battlespace enable the UMT to provide comprehensive religious support despite the confusion and chaos of battle. Coordination with key staff members and the executive officer is critical.

> The UMT must employ sleep plans and staggered shifts when necessary to remain fully capable of night operations.

4-127. Task force operations are continuous. The UMT must be fully capable of night operations, employing sleep plans and staggered shifts when necessary. If the TF moves during the hours of darkness, the TF UMT must maintain contact with other elements with which it is traveling.

4-128. The attacking force moves to positions of advantage during this time. The basic forms of maneuver are envelopment, penetration, frontal attack, turning movement, and infiltration.

4-129. In the maneuver phase, religious support may be restricted to small numbers of soldiers during halts in the movement, or at casualty collection points along the route of march.

4-130. During the assault, casualties are likely to increase. The TF UMT will be heavily engaged with providing religious support to casualties and caregivers. If overwhelmed with casualties, the BAS declares a MASCAL situation. The TF UMT requests additional support during a MASCAL from the brigade UMT.

4-131. During a MASCAL the TF UMT requests additional support from the brigade unit ministry team.

Employ Firepower

4-132. The fire support coordinator (FSCOORD) integrates fire support into the unit's maneuver scheme. Religious support during this phase is similar to the maneuver operation.

Protect the Force

4-133. To protect the force a commander can employ any number of actions to include: maintaining a high tempo of operations; conducting security operations, which include organizing and employing covering, guard, or screen forces; conduct area security operations; employ operations security (OPSEC) procedures; execute deception operations; conduct defensive information operations; employ camouflage, cover, and concealment, and conduct active and passive air defense operations.

4-134. The UMTs movement in the battlespace, communication methods, and information management techniques should reflect the commander's security and force protection objectives. The UMT closely monitors force security issues that may affect religious support planning The UMT aggressively monitors the battle and the commander's use of command and control (C2) in allocating forces. The decision-making process and the format for plans and orders are found in FM 101-5. Below are the UMTs planning considerations within this staff process.

The Goal: Comprehensive/Synchronized Religious Support		
Staff Action	***UMT Action for input to staff***	***UMT Action for internal use***
Receive the Mission	-Mission Analysis: Can we support the mission and how? -Coordinate with S4 and S1. -Coordinate with higher UMTs.	-Begin rehearsals of known actions/drills. -Execute current RS plan. -Begin considering the unit's proposed COA's for support of the mission.
Conduct Estimates by each staff section/	-Prepare RS Estimate of current/projected unit condition. -Provide input to staff on AOR religions and culture.	-Communicate issues to subunit UMTs. -Continue Spiritual Care.
Restated mission to Cdr/ Receive Initial Cdr's Guidance	-Review religious, moral, and ethical issues.	-Continue prep for next mission.
Develop/compare/ recommend the 'best' COA to Cdr	-Input to COA's for tactical mission. -RS Implications of each COA. -Compare feasibility/ acceptability/ supportability of each COA that the staff produces.	-Further develop RS plan for support of the upcoming mission. -How will we synchronize RS? -Continue provide RS to unit and area personnel.
Receive Cdr's Estimate including DECISION and Cdr's CONCEPT	-Refine RS concept to comply with Cdr's concept.	-Communicate with subunit UMTs concerning their own preparation of subunit RSP needs and issues.
Prepare Plans and Orders	-Submit RS annex/paragraph if needed.	-Update RS Section NCOIC. -Continue RS.
Receive Cdr's approval of Plans and Orders	-Pray with Cdr, and give 1 min update of RS situation.	-Continue information gathering. -Continue RS.
Issue Plans and Orders	-Attend OPLAN meeting; be prepared to brief.	-Communicate RS plan and provide support to higher RMSTs and subunit UMTs.
Rehearsals and Troop Leading Procedures	-Attend rehearsals, insure RSP is integrated and understood.	-Aggressively execute RS plan.
Conduct mission	-Begin planning for follow on mission, sequels, branches.	-Provide prioritized RS. -Battle track subunit UMTs.

4-135. The UMT must remain flexible and support changes to the plan due to METT-TC considerations and the commander's priorities.

PERFORM LOGISTICS AND COMBAT SERVICE SUPPORT

4-136. On reaching the objective, the unit eliminates remaining resistance, consolidates its gains, and reorganizes as needed. The unit may immediately continue the attack to a follow-on objective, or it may hold in place and set up a hasty defense.

> The UMT provides religious support to the force paying attention to leaders and those who show signs of battle stress.

4-137. The time the TF spends during this part of the offensive operation will depend upon subsequent missions. If there is sufficient time, the UMT moves forward to maneuver elements. The UMT provides religious support to the force paying special attention to leaders and those who show signs of battle stress. The team may also conduct memorial services or ceremonies as directed by command.

4-138. A successful assault may develop into an exploitation or pursuit. The effectiveness of an exploitation or pursuit depends on rapid transition and vigorous execution. The UMT must be prepared to make the transition quickly with the least disruption to its operations.

TRAVEL IN THE BATTLESPACE

4-139. A UMT that travels to forward elements by itself is at great risk. Travel with the TF LOGPAC convoy reduces the security risk and the chances for getting lost. The LOGPAC moves from the field trains along the MSR to an LRP where the company first sergeants take control of their company LOGPACs.

> The UMT meets with company first sergeants at the LRP.

4-140. Based on the scheme of maneuver, the TF S4 selects LRPs well forward. For security reasons, they are only used for short periods of time. The UMT finds LRP times and locations in the OPORD.

4-141. The S4 and company first sergeants meet at the LRP to discuss future logistical requirements. At this meeting, the UMT discusses the religious support needs of soldiers and coordinates for services and visits to the line companies with the first sergeants. The UMT then may go to one of the companies with the first sergeant and company LOGPAC

4-142. When the company has completed its re-supply, the first sergeant returns his portion of the LOGPAC convoy to the LRP. The TF support platoon leader reforms the convoy and leads it back to the field trains. The TF UMT follows the first sergeant to the LRP and returns to the combat trains or, if time permits, travels to the field trains.

4-143. Engineers. The engineers provide an alternative to traveling with the LOGPAC. Because LOGPACs are often conducted during times of limited visibility, the UMT may find travel with elements of the engineer company more effective. The engineers travel to all the company locations to dig in hasty defensive positions, clear mines and obstacles, and prepare for the follow-on mission. The UMT coordinates with the S4 to link up with the engineers for travel to the companies.

RELIGIOUS SUPPORT OF DEFENSIVE OPERATIONS

4-144. Army forces conduct defensive operations as part of major operations and campaigns in combination with offensive operations. The basic task of the defense is to force the enemy to attack under unfavorable circumstances. Military forces defend only until they gain sufficient strength to attack. Though the outcome of decisive combat derives from offensive operations, it is often necessary, even advisable, to defend.

4-145. Defensive planning considerations are very similar to offensive considerations. In the defense, the UMT has more time to synchronize the religious support plan and apply maximum religious support assets to critical areas. Shifting religious support assets as the operation evolves will require prior coordination with commanders and UMTs. Prior planning regarding the synchronization of the religious support mission and plan during different phases increases the effectiveness of the religious support defense plan.

BATTALION/ TASK FORCE

Develop Intelligence

4-146. As soldiers prepare defensive positions, the TF executes the counter-reconnaissance mission. This mission is meant to deny the enemy information about friendly dispositions and friendly preparations for battle. The counter-recon force typically consists of the scout platoon, ground surveillance radars, and maneuver elements.

> Starting as far forward as possible, while minimizing risk to exposing soldiers to artillery fire, the UMT makes brief visits to soldiers in fighting positions.

4-147. Throughout this phase soldiers are digging in and preparing fighting positions for the coming enemy attack. The UMT keeps movement to a minimum to avoid disclosing fighting positions, paying close attention to cover and concealment. Starting, as far forward as possible, the team makes brief visits to soldiers in fighting positions, working rearward towards the combat trains. The UMT coordinates its travel with the TOC and the combat trains command post (CTCP).

> The UMT must carefully plan movement and religious support to avoid being caught in the confusion.

Deploy/Conduct Maneuver

4-148. The rearward passage of lines by the covering force causes considerable movement and confusion. The UMT must be cautious when planning movement and religious support to avoid being caught in the confusion. The UMT locates with the battalion aid station or LOGPAC, communicates its location to the TOC and CTCP, and deliberately avoids the passage lanes.

Employ Firepower

4-149. Enemy artillery poses a significant threat to the safety of the UMT. The UMT must be near adequate overhead cover for protection. Gathering soldiers for worship may involve undue risk.

4-150. If the enemy uses chemical or biological agents, the team will make adjustments in religious support. The team may need to postpone religious activities until the hazard is eliminated or reduced to a safe operating level. Several factors that the UMT considers include:

- Degree of engagement with enemy forces.
- Level of exposure and risk to soldiers and equipment.
- Priorities for religious support.
- Anticipated operations in response to enemy attack.

The chaplain offers sacraments and ordinances to soldiers before combat.

4-151. Sacraments and ordinances are not celebrated in contaminated areas. The UMT does not resume services until soldiers have moved to safe areas. Therefore, the chaplain takes advantage of every opportunity to administer the sacraments before soldiers engage in combat, or before soldiers are exposed to a contaminated environment.

4-152. The UMT carefully coordinates movement around and through obstacles set in place for the defense. Obstacles installed after the team has departed its base of operations can put the team at risk. The team should know the breach marking system and MSR. To avoid obstacles, it carries a copy of the obstacle overlay.

The team carries a copy of the obstacle overlay when traveling.

4-153. The UMT must be prepared to adapt to the changing tactical situation of a mobile defense that combines offensive, defensive, and retrograde actions. This combination results in a nonlinear front to create confusion for the attacking forces.

4-154. As the enemy begins the attack, the UMT shifts its attention to soldiers in the battalion aid station (BAS). From this point on, the UMT's main effort will be the care of dying and other casualties as they are brought to the BAS.

The UMT must be prepared to move on a moment's notice.

4-155. During the enemy attack, the BAS may move frequently, evacuation routes may be blocked by an enemy penetration, and defending elements may become temporarily encircled or bypassed by enemy forces. Rapidly moving enemy units may threaten or overrun the BAS. The UMT must be prepared to move on a moment's notice. Equipment and supplies not required immediately should be stowed in the team's vehicle. The vehicle should be parked in a manner that allows for rapid departure if the BAS is threatened.

Perform Logistics and Combat Service Support

4-156. As the TF consolidates and reorganizes after the battle, the companies evacuate the remaining casualties to the aid station. Once the UMT has ministered to the dying and other casualties, it assesses METT- TC to determine its next actions. If the unit is planning a counterattack, the UMT prepares to move with the unit. If not, the team moves forward to units on the forward line of own troops (FLOT) to visit soldiers. The UMT NCOIC ensures the team's supplies are adequate for follow-on operations.

Coordinate with Command

4-157. The UMT must have a clear understanding of the battlespace situation so they can prioritize tasks at the right time and location. The UMT takes advantage of the war-gaming that takes place in the MDMP to derive their decision points. The UMT coordinates with the commanders and other UMTs in the AO to ensure all religious support needs are met according to the operational plans.

BRIGADE

4-158. In the mobile defense, battalions engaged in offensive, defensive, and retrograde actions present a challenge for the brigade UMT. The UMT plans religious support for each of the units based on METT- TC. The area defense differs from the mobile defense in that all units are engaged in defensive operations. Within the defense planning considerations by the staff, the brigade UMTs responsibilities to the BDE UMTs include—

- Faith Group support.
- Operational information.
- Logistics of Faith Group specific supplies.
- Personal spiritual support to UMTs.

STABILITY OPERATIONS

4-159. Conducting religious support in stability operations is essentially the same as in offense, defense, and support operations.

Battalion

4-160. The UMT will need to reprioritize planning considerations in order to fully support this type of operation. The UMT planning emphasis will include—

- Joint and multinational religious support cooperation and communication.
- More planning for faith-based NGO/IO liaison tasks.
- Working more closely with the CMO personnel in mutual support tasks.
- Host nation religious leadership liaison responsibilities.

Brigade

4-161. The BDE or equivalent will be challenged with coordination and information management requirements. The information flow will require the proper assessment of information to allow for timely and adequate religious support. Planning considerations in addition to BN include—

- Civil Military Operations Center (CMOC) representation involvement.
- Staff integration needs.
- Media and public affairs command guidance.
- Remote site religious support coverage planning.
- Force protection.
- Religious support to Contractors.

SUPPORT OPERATIONS

4-162. The tasks for planning religious support for Support Operations are identical for offense, defense and stability operations. The support operation requires the UMT to know the restrictions, limitations, and proper responsibilities of the UMT in providing religious support in a support operation environment.

Battalion

4-163. The UMT will be directly involved with civilians and the trauma associated with the operation. This will have a stressful effect on the UMT members. The BN UMT must be prepared for the area of operation METT-TC considerations. Planning includes—

- Personal burnout and combat stress.
- A daily log (See DA Form 1594).
- Soldier and civilian debriefings.
- Continual spiritual and morale assessment of individuals in the unit.

Brigade

4-164. The BDE UMT will work closely with state agencies and joint chains of command. The religious support plan and operation must operate within the mission, concept of operation, and commander's intent. Legal restraints may hinder the free exercise of religion as it relates to the military providing religious support to civilians who request assistance and guidance. The UMTs battle book and TSOP must reflect the planning considerations. Planning considerations include—

- Mass Casualty training.
- Information on civilian and state agencies for liaison tasks.
- Critical Event Debriefing as a norm.
- Moral and ethical dimensions.

TERMINATION AND POST-CONFLICT OPERATIONS

4-165. Planning for termination of operations must be ongoing during all phases of COA development, deployment of forces, and execution of operations. It must be fully integrated within the orders process and synchronized and coordinated with the staff and UMTs. In most cases, operations will be terminated in their own right when stated objectives have been met. In some operations (e.g. stability and support), the command may have to recognize that the mission is not achievable without restructuring and/or commitment of additional assets. Termination of the mission may be the best COA. The following checklist may assist in termination planning:

- What is the planned timeline for termination?
- Will there be follow-on forces and who?
- What are the post-conflict requirements?
- Is there a force protection issue for safely withdrawing forces?
- What are the requirements for redeployment?
- What are the considerations for release of the Reserve Component forces?

- Will there be humanitarian, governmental, and infrastructure assistance required for post conflict activities? Is there coordination required?

(See JP 5-00.2 and JP 1-05 for more detail.)

TRANSITION

4-166. As with planning for conflict termination, planning for transition must extend throughout the planning process and into operations and redeployment. Transferring control of command and staff responsibilities to another organization requires detailed planning and execution. It is dependent on the situation and each transfer will have it's own characteristics and requirements. Transition planning by the UMT continually requires close coordination with higher and communication with the UMTs of the transitioning organization. Regeneration and reconstitution is a primary consideration in the transition plan. Transition planning by BDE should encompass the following:

- What is the new mission?
- What U.S. forces, equipment, and/or supplies will remain behind?
- Who will support UMTs that remain behind?
- What is the classification of information to be shared with incoming UMTs?
- What will be the new ROE?
- What are the building/Chapels requirements?
- What is the timeline for transfer?

The mission may dictate that additional planning may considerations need to be planned and coordinated for as well.

REDEPLOYMENT AND RECONSTITUTION

4-167. Reconstitution (see FM 100-9) is a method of replenishing units that have been decimated to the extent they are no longer combat-effective. Such units are withdrawn from contact to a rear area. Personnel and equipment as well as rest and recuperation are the focus. These units require special religious support. BDE UMTs must be proactive in supporting reconstitution not only for unit personnel, but the UMTs of that unit, its leadership, and caregivers as well. Religious support will encompass the following:

- Honoring the dead with memorial services and ceremonies.
- Nurture of the living with counseling, worship, and critical event debriefings.

4-168. Post-conflict requirements and impact have a direct bearing on the redeployment religious support mission. The impact of battle and the stress associated with personal loss, fear, physical and emotional fatigue demands religious support application. METT-TC is still utilized in this stage by UMTs as it was in deployment. The balancing of available assets is essential during this phase to ensure effective religious support. UMTs must factor in redeployment not only back to home station, but also to other missions. Both require extensive religious support planning. UMTs of the BN, BDE, and equivalents must not re-deploy prior to their completing their religious support mission to their respective units and other associated religious support requirements of the TF. Family support and other religious support missions at home station is the responsibility of the installation religious support mission. Planning considerations for both echelons include—

- After Action Report.
- Reunion Briefings program.
- Logistics and time table.
- Religious support coverage plan.
- Echelon of UMTs back to home station ensuring AO religious support coverage.

GARRISON OPERATIONS

4-169. UMTs participate in the religious support program of the installation in accordance with their orders of assignment by Headquarters Department of the Army. They coordinate with their chain of command and the installation chaplain for garrison religious support needs for their unit soldiers and families while non-deployed and during all phases of deployment. Planning considerations include—

- Family Readiness Group (FRG): Support by briefings/classes on family issues.
- Killed in action (KIA)/missing in action (MIA)/wounded in action (WIA) memorial ceremonies and religious services.
- Hospital visitation.
- Casualty notification support.
- Worship support.

OTHER BRIGADE AND BELOW UNITS AND SPECIAL RELIGIOUS SUPPORT CONSIDERATIONS

Special Forces

4-170. Special Forces (SF) combine at the lowest tactical level, the functions performed by conventional branches of the Army. United States Army Special Operations Command (Airborne) (USASOC (A)) is the Army component of the joint and unified command, United States Special Operations Command (USSOCOM). USASOC (A) is comprised of Special Forces, Rangers, a Special Operations Aviation Regiment, the Special Operations Support Command and Civil Affairs and Psychological Operations forces. Special Forces is commanded by the U.S Army Special Forces Command (USASFC), which has seven subordinate commands known as Special Forces Groups consisting of three Special Forces (SF) battalions each. USASFC (A) has five primary missions and related additional capabilities.

Primary Missions

4-171. The five primary missions are Unconventional Warfare (UW), Foreign Internal Defense (FID), Special Reconnaissance (SR), Direct Action (DA), and Counter Terrorism (CT). An SF unit can have elements deployed to several different areas within its assigned region. These elements range in size from the individual soldier to units of 12 to 150 soldiers. SF missions are directly influenced by politico-military considerations. A change in national security policy or strategy may change SF missions.

Additional Capabilities

4-172. Special Forces units are trained to conduct MOOTW missions. Language capabilities, area orientation, small unit, and multi-agency experience are invaluable in MOOTW. Additional capabilities include noncombatant evacuation operations (NEOs), arms control, support to domestic authorities, humanitarian assistance (HA), disaster relief, security assistance, nation assistance, counter-drug operations, combating terrorism, peacekeeping and peace enforcement, show of force, support to insurgencies and counterinsurgencies, personnel recovery, and special activities.

4-173. The Special Forces branch or related MOS is the only non-accessioning branch or MOS. A soldier selected for the Special Forces branch or MOS has successfully completed service in an accession branch or MOS and graduated from the Special Forces Qualification Course.

UMT Personnel Qualifications

4-174. Religious support in the Special Forces community requires unique qualifications. The most critical is the ability to provide sustained, independent religious support. Others include—

- Security clearance for all UMT members equal to the mission classification.
- Airborne qualification.
- Extensive orientation in special operations, or graduation from the Special Forces Qualification Course, or Special Forces Staff Course (SFSC), or an "S" enlisted MOS identifier.
- Training in the religious background of the region of deployment.

The Nature of Religious Support in Special Forces Units

4-175. The unconventional nature of special operations, the high OPTEMPO, and the multiple mission deployments create stress on units, soldiers, and families. These pressures place special demands for religious support on SF UMTs.

4-176. The USASFC (A) chaplain, under the guidance of the USASOC (A) Chaplain, gives staff supervision to the USASFC (A) religious program. Group chaplains supervise religious support operations and provide technical guidance and resources to subordinate UMTs. SF chaplains frequently deploy to advanced operational bases (AOB). If a Special Forces Operational Base (SFOB) is deployed, a Catholic chaplain, when available, is attached to perform religious support at the SFOB, as well as, at the AOB, and down to the Operational Detachment Alpha (ODA).

Ranger Regiment

4-177. The 75th Ranger Regiment is a major subordinate command (MSC) of the USASOC (A). The Ranger Regiment is capable of conducting both conventional and special operations across the full spectrum of conflict. As a potent, rapidly deployable direct action infantry force, it is particularly well suited to conduct short duration strike operations or other small-scale offensive actions against targets of high strategic or operational importance. Such operations include actions to seize, destroy, or capture enemy forces or facilities, or to recover designated personnel (noncombatant evacuation operations, liberate friendly prisoners of war, capture designated enemy personnel) or equipment from hostile, denied, or politically sensitive, high threat environments requiring exceptionally competent, mature and disciplined forces.

4-178. The Rangers may conduct operations independently, or in support of a campaign plan. While Ranger operations may be conducted in coordination with conventional forces, they differ from conventional operations in degree of risk, operational techniques, and modes of employment. Ranger operations will on occasion be conducted before political preparation of the battlespace has been completed or, in certain cases, before such preparation has even been initiated. Because of its rapid deployment ability and capabilities as a formidable strike force for forced entry operations; the Ranger Regiment maintains an exceptionally high level of operational readiness.

Organization

4-179. The Ranger Regiment's structure is similar to that of a conventional light infantry brigade, but with a robust regimental headquarters that includes organic reconnaissance, signal, military intelligence, and training detachments. Ranger units are not designed for sustained combat and therefore, they are limited by very austere organic logistic support, and by equally limited organic combat multipliers. During all phases of operations and training, Ranger units require responsive and adequate support, either from theater Army assets or USASOC (A).

Unit Ministry Team Personnel Qualifications

4-180. Religious support to Army Rangers is physically, emotionally and spiritually demanding. Unique qualifications include—

- Security clearance for all UMT members equal to mission classification.
- Airborne qualification.
- Ranger qualification.
- Completion of the Ranger Orientation Program (ROP) by chaplains and assistants E-5 and above.
- Completion of the Ranger Indoctrination Program (RIP) by chaplain assistants E-1 through E-4.

The Nature of Religious Support in Ranger Units

4-181. All Ranger elements, from the single platoon to the entire Ranger Regiment, are capable of conducting extremely decentralized training exercises and combat operations. Some additional religious support planning requirements include:

- A thorough understanding of the operational and strategic mission.
- Crisis action planning and execution of religious support

4-182. The Ranger Regimental chaplain exercises staff supervision over regimental religious support and subordinate Ranger battalion UMTs.

U.S. ARMY CIVIL AFFAIRS AND PSYCHOLOGICAL OPERATIONS COMMAND (AIRBORNE)

4-183. U.S. Army Civil Affairs and Psychological Operations Command (Airborne) (USACAPOC (A)) provides civil affairs (CA) and psychological operations (PSYOP) activities and information for both SF and conventional commanders. The CA/PSYOP mission is to inform commanders on the nature of foreign audiences to influence the behavior of foreign governments, organizations, groups, and individuals. To influence a target audience's thinking and behavior, the CA/PSYOP unit must be extremely well versed in the language, culture, thought patterns, and belief systems of the people in the AO.

PSYCHOLOGICAL OPERATIONS UNITS

Organization

4-184. Psychological Operations (PSYOP) personnel often deploy for extended periods as individuals and small teams to support SOF and conventional forces. Deployed personnel are required to work with high-level leaders of other armed forces, and with government officials from the U.S. and other nations.

Psychological Operations Unit Ministry Team Personnel Qualifications

4-185. Requirements for members of UMTs in PSYOPS varies between components to include:

- Security clearance for all UMT members equal to the mission classification.
- Airborne Qualification.

The Nature of Religious Support in Psychological Operations Units

4-186. Tailored PSYOP units in support of SF and conventional forces require flexible religious support. UMTs deploy with a battalion-size or larger PSYOP Task Force (POTF) or PSYOP Task Group (POTG). When POTF or POTGs do not deploy, the religious support is coordinated through the Supported unit.

- PSYOP UMTs frequently coordinate religious support with joint or combined forces.
- When a PSYOP UMT does not deploy, the supported unit provides religious support to PSYOP personnel.
- PSYOP UMTs participate extensively in family support programs during all stages of deployment.

CIVIL AFFAIRS UNITS

4-187. Civil Affairs (CA) units support the commander's relationship with civil authorities and the populace, promotes mission legitimacy, and enhances military effectiveness. CA units support all USASOC (A) missions with regional orientation briefings and information on political, social, cultural, religious, and economic aspects of the Joint Special Operations Area (JSOA).

Organization

4-188. CA units are task organized to support USASOC (A) and supported conventional units' mission requirements. Some CA units are organized as battalions with FID and UW support missions. Other units are organized into special teams, such as Cultural Affairs teams.

Unit Ministry Team Personnel Qualifications

4-189. The assignment of a Unit Ministry Team requires unique qualifications, to include—

- Security clearance for all UMT members equal to the mission classification.
- Airborne qualification.
- Extensive knowledge of religious and cultural issues and their effect on operations.

The Nature of Religious Support in Civil Affairs Units.

4-190. UMTs serving within USACAPOC (A) primary missions are to provide and perform religious support to the unit assigned in accordance with U.S. Code Title X. Civil Affairs UMTs may provide all UMTs in an AO a unique avenue of information and expertise on religious cultures in the AO. CA UMTs can also coordinate for command directed CA activities involving liaison between the command, local religious leaders, and UMTs. CA teams attached to supported commands are provided religious support through the supported command UMT. CA UMTs serve as another religious support asset for the combatant command chaplain's religious support plan.

MEDICAL UNITS

4-191. Deployable Army Medical units are in the process of transformation under the Medical Reengineering Initiative (MRI). MRI is expected to be complete by FY 2009.

4-192. Under MRI, deployable medicine includes C2 units, corps medical augmentation units, and hospitals. MRI C2 units include Medical Brigades, and Medical Commands (MEDCOMs) at Corps and Echelons Above Corps. MRI hospitals come in one type: the Combat Support Hospital, which can be located at Corps and Echelons Above Corps.

Organization

4-193. Unit Ministry Team personnel are organically assigned to MRI C2 units and to MRI hospitals. Corps medical augmentation units receive their religious support through area coverage.

4-194. Since health care skills are highly perishable and in great demand, the largest number of the Army's health care providers are assigned to TDA Medical Treatment Facilities (MTFs). When needed for operational requirements in deployable units, these health care providers fill necessary slots in deploying MTOE medical units under the Army Medical Department (AMEDD) Professional Officer Filler System (PROFIS).

4-195. Likewise, since medical ministry and pastoral care skills are also highly perishable and in great demand, Unit Ministry Team personnel may be assigned to TDA MTFs with PROFIS obligations to deployable MTOE medical units.

Unit Ministry Team Personnel Qualifications

4-196. Qualifications for Unit Ministry Team personnel serving medical units include:

- 7S, Combat Medical Ministry Specialist and
- 7R, Hospital Ministries Clinician for Chaplains;
- Emergency Medical Ministry Course (EMM) for chaplain assistants.

4-197. To meet the complex demands of combat medical ministry and pastoral care, the AMEDD Center and School's Chaplain Training and Operations Branch oversees an institutional training program to prepare Unit Ministry Team personnel providing religious support in medical unit. This program includes—

4-198. Emergency Medical Ministry Course (EMM) is the entry-level course, training chaplains and chaplain assistants in the basics of wartime emergency medical ministry. It is a two-week TDY course taught at AMEDD Center and School, with practical application at Brooke Army Medical Center. EMM is the basic qualifying course for Chaplain Assistants serving in medical units.

4-199. Combat Medical Ministry Course (CMM) is the intermediate course and requires one prerequisite unit of accredited Clinical Pastoral Education (CPE). Also a two-week TDY course taught at AMEDD Center and School, with practical application at Brooke Army Medical Center. CMM focuses upon six core competences: Crisis Ministry; Trauma Ministry; Mass Casualty Ministry; Biomedical and Battlespace Ethics; Grief, Loss, Death and Dying; and Spiritual Hope/Suicide Prevention. CMM is the qualifying course for award of the 7S, Combat Medical Specialist, ASI, the minimum qualification for Chaplains serving medical units.

4-200. Clinical Pastoral Education (CPE): Residency is the advanced level, accredited by the Association for Clinical Pastoral Education, Inc. and administered by the MEDCOM Army CPE System. The Army CPE system offers residencies at Walter Reed, Eisenhower, Madigan and Brooke Army Medical Centers. CPE prepares Chaplains for service within both institutional and deployed medical environments. CPE is required for award of the 7R ASI.

4-201. Clinical Pastoral Education: 7R Trainer (CPE 7R) is the fellowship level. Accredited by CPE, CPE 7R prepares select 7R Chaplains to supervise CPE residency training within the Army CPE System. CPE 7R is required for award of the 56D7R AOC.ASI.

The Nature of Religious Support in Medical Units

4-202. Unit Ministry Teams serving with medical units bring specialized training in spiritual assessment: mass casualty ministry; crisis ministry; grief and loss; spiritual hope; trauma ministry; and ethics to their provision of religious support. They have a dual religious support focus, in caring for the dying and other casualties and honoring the dead, while also nurturing the living in support of the medical staff.

SUMMARY

4-203. The UMT is assigned to a unit at either BN/Squadron or BDE/Regiment/ or Group. The unit is either operationally committed, about to be deployed, returning from deployment, or is in a garrison situation. In garrison, the priorities of religious support must fit in with the command's primary responsibilities:

- Training.
- Family Support.

4-204. Key to performing religious support in this environment is executing staff responsibilities to fulfill various tasks assigned to the UMT in garrison.

4-205. For the operationally committed unit, the UMT is involved in the highest calling for military support: supporting the soldier in combat situations, operations that fall into four categories: Offensive, Defensive, Stability and Support.

4-206. UMTs supporting these units may participate in joint command, and multi-national operations.

RESOURCES

4-207. The following is a list of main resources utilized for writing and updating this doctrine.

- AR 165-1.
- JP 1-05.
- JP 5-00.2.
- JP 3-05.
- FM 16-1.
- FM 3-0 (100-5).
- FM 3-90.3 (71-3).
- FM 4-0.
- FM 25-100.
- FM 25-101.
- FM 71-2.
- FM 100-9.
- FM 100-10-2.
- FM 100-21.
- FM 7-30.
- FM 7-20.
- FM 1-111.
- Lessons Learned from past and current operation
- Lessons Learned from the training centers

Chapter 5

DIVISION UNIT MINISTRY TEAM RELIGIOUS SUPPORT

TACTICAL ECHELON DESCRIPTION

CONTENTS

5-1. The division is the largest fixed Army organization that trains and fights as a tactical team. The division echelon possesses great flexibility. The division is a unit of maneuver, organized with varying numbers and types of combat, combat support (CS), and combat service support (CSS) units. It may be armored, mechanized, medium, light infantry, airborne, or air assault. Each are capable of conducting operations over a wide range of environments. Their commanders tailor their organic brigades and attached forces for specific combat missions.

5-2. The early 1990s brought significant changes to the world and the Army. The end of the Cold War presented threats that were more ambiguous and regionally focused. Force Projection has become our strategy for supporting the power-projection element of U.S. national security. Army divisions are prepared for force projection operations, as part of a joint, interagency, or multinational effort for full spectrum operations.

5-3. U.S. Army corps and divisions conduct Army operations that span the spectrum from peaceful engagement to actual combat. Army divisions train to fight and win battles and engagements as part of a joint force anywhere in the world. Their organization, capability, flexibility, and versatility allow them to conduct operations for stability and support multinational and interagency operations, as well as, in joint operations. The unique capabilities of armored, mechanized, infantry, light infantry, airborne, and air assault divisions provide the U.S. Army the ability to serve the national interests worldwide. (See FM 71-100 for further information.)

AS PART OF A CORPS

5-4. Traditionally, divisions have operated as part of a U.S. corps. In corps operations, divisions normally comprise 9 to 12 maneuver battalions, organic artillery battalions, and supporting CS and CSS units. Divisions perform a wide range of tactical missions and, for limited periods, are self-sustaining.

Corps augments divisions, as the mission requires. A corps or a numbered army may direct a division to perform tasks of operational importance.

AS PART OF A JOINT TASK FORCE

5-5. Division decision-making processes and capabilities make Army divisions effective in joint, multinational, and interagency operations. Army divisions focus on fighting and winning battles and engagements as part of a joint force anywhere in the world.

5-6. The division may deploy as part of a JTF without its traditional corps headquarters and corps support unit. In these types of operations, the division often works directly for the JTF commander. Therefore, divisions must know joint doctrine, tactics, techniques, and procedures. Divisions assigned to a JTF normally conduct traditional tactical operations and may be involved in nontraditional actions, such as interagency operations or operations with the host nation. Divisions are not normally designated as a JTF headquarters.

5-7. A division in a JTF requires experienced liaison personnel. The division may operate with or receive support from joint, multinational, interagency, or NGOs. The division may receive mission support such as joint fires, joint air defense, national level intelligence, and theater and national level communications. Competent liaison to coordinate with these agencies and other headquarters helps ensure properly executed operations.

JOINT TASK FORCE STAFF AUGMENTATION

5-8. Without significant augmentation, a division headquarters has a limited capability to perform as an Army force (ARFOR) or JTF.

DIVISION AS AN ARMY FORCE

5-9. Depending on the mission assigned to the JTF, the Army forces required for the operation may only be a division or a tailored force of units from a division. The division may find that it is the largest land component command group or staff organization assigned to the JTF. In this scenario, the JTF commander may designate a division under its C2 as the Army Forces (ARFOR) headquarters in small-scale contingency (SSC) operations, i.e., the senior Army headquarters within the JTF. Operations include foreign humanitarian assistance, peace operations, noncombatant evacuations, enforcing sanctions, reinforcing key allies, and limited strikes and interventions. Operations are conducted with and in support of government agencies, nongovernmental organizations, regional and international security organizations, and coalition partners. A division designated as an ARFOR executes these responsibilities unless they are specifically excluded. The Army Service Component Command (ASCC), in coordination with the division and DA staff, tailors the forces provided to the ARFOR to meet specific METT-TC conditions anticipated in the JOA. The ARFOR UMT must plan additional RS assets for unit, area, and denominational coverage.

5-10. Designation of the division as the Joint Forces Land Component (JFLC) or the ARFOR of the JTF incurs numerous additional staffing responsibilities. The division will have to coordinate with other JTF-level component or functional commands as well as the service component headquarters of the

JTF establishing authority for administrative and logistic support. The division as ARFOR plans and conducts operations in accordance with the CJTF guidance and detailed plans. The ARFOR will establish liaison personnel to the CJTF, other component commanders, and supporting commanders as necessary or as directed by the CJTF. The division staff will require augmentation to its staff to accomplish the tasks inherent to operating as the JFLC or ARFOR. In the event that a deployed division becomes the JFLC or ARFOR for the JTF, the commander must request staff, CS, and CSS augmentation. The division may need assistance in providing the necessary support units and sustainment activities for prolonged support of significant numbers of assigned or attached forces. (See FM 3-91 (71-100), Appendix F for further information.)

FUNDAMENTALS OF RELIGIOUS SUPPORT OPERATIONS

5-11. The mission of the Division UMT is to coordinate religious support to soldiers, families, and as the force structure evolves, to forward-deployed authorized civilians. Division Chaplains also serve as critical advisors to commanders at all levels of the division providing essential information on soldier and unit morale as affected by religion, quality of life matters, moral/ethical decision making, and the impact of indigenous religion on military operations. The Division Chaplain develops the CMRP and the division RSP on behalf of the Division Commander.

5-12. The battlespace of the 2lst century involves conditions of modularity, split-based operations, increased lethality, non-linearity, greater troop dispersion, increased agility, increased pace of operations, sustainment of tempo and simultaneity of combat operations.

5-13. Combat power in the 21st century will rely heavily upon technology. The cumulative affect of the numerous factors will increase the spiritual, psychological and interpersonal demands placed upon soldiers of the 21st century. Technology can provide additional means to plan, coordinate, and execute religious support to address these demands, but the personal delivery of religious support by the UMT will always be imperative.

ARMY FORCES

5-14. As an Army forces (ARFOR), the impact upon religious support planning and execution is significant. The role of the ARFOR UMT is significantly more diverse and broad than a division operating as a tactical headquarters in a corps. While the division plans, prepares for, and executes many of the same types of activities, its focus is always on the tactical operation in its AO. Unlike the division UMT focus, the ARFOR UMT not only focuses on tactical religious support tasks in its AO, but also provides critical operational-level religious support sustainment to U.S. Forces UMTs and others throughout the JOA. The ARFOR UMT may provide religious support management to tactical and sustainment activities in contiguous or noncontiguous areas, or a combination of both. The ARFOR UMT must focus its planning on multiple, simultaneous actions that accomplish its own, or sustains other's, religious support mission. The tasks performed include—

- Information management including processing, assessment, and dissemination for command and subordinate UMTs from strategic and operational sources.
- Managing religious support tactical missions and tasks to accomplish missions assigned by the Joint Force Commander or Joint Force Land Component Commander.
- Planning and executing self-deployment of religious support assets to ensure comprehensive and seamless religious support.
- Planning and integrating CMO mutual support tasks.
- Coordinating with Special Operations Forces (SOF) UMTs.
- Executing religious support sustainment operations.
- Providing logistical support.

JOINT TASK FORCE

5-15. The joint task force (JTF) chaplain is responsible for planning, coordinating, and implementing all religious support within the JTF for the commander of the JTF. In addition to the already established doctrinal tasks, the JTF UMT is responsible for the following:

- Developing joint religious ministry support plans in support of the operation. This includes preparation of the Chaplain Activities Appendix, under the Personnel Annex of the CJTF's OPLANs, OPORDs, or campaign plans, to include appropriate tabs.
- Identifying RMST requirements such as augmentation, logistics, and communication requirements.
- Coordinating religious ministry with MSC Chaplains.
- Maintain liaison with chaplains of international forces and appropriate host nation (HN) civilian religious leaders.
- Coordinating appropriate training for RMSTs.
- Coordinating with the other staff sections to ensure sufficient religious assets are available to support combat forces, medical services, civilian detainees, EPWs, and others.
- When appropriate and in coordination with the CMOC, the JTF may conduct liaison with faith-based NGOs and IOs.
- Advising distribution of humanitarian assistance supplies from religious organizations.

PLANNING CHECKLIST

- General
  1. Staff: communication, annex, integration, briefings
  2. Specialized:
- Nurture the living:
  - Spiritual Fitness
  - Worship
  - Counseling
  - Visitation to staff and subordinate UMTs
- Care for the Dying:
  - Spiritual triage
  - Leadership
  - Battle Stress
- Honor the Dead:
  - Memorial Services
  - Memorial Ceremonies
  - Rites
  - Sacraments
  - Prayers

Predeployment

5-16. The Division UMT translates the commander's intent and decisions into a fully coordinated and supported religious support operation. Division UMTs, in coordination with their commander and staff, can—

- Accurately determine the current situation.
- Meaningfully assess courses of action for the commander.
- Issue the necessary guidance and instructions, having received guidance from the chief of staff.
- Monitor how well the commander's religious support plan and instructions are being executed.
- Alter the plan in accordance with unit feedback, commander's guidance, commander's intent, and METT-TC.
- Plan future operations.
- Synchronize actions prior to combat operations.

5-17. The division UMT is expected to operate with a great deal of freedom and responsibility. From information available, it must discern what is essential for the commander's decision making. It must coordinate the entire division religious support effort to support the commander's intent and concept of operation. Normally, the commander issues guidance and makes tactical decisions from field locations, then moves to the point of the division's main effort. The UMT develops appendices and plans necessary coordination between subordinate and higher UMTs to execute the commander's intent.

Deployment

5-18. Religious support, like other battlespace support systems, requires C4I and leadership. The Division Chaplain provides UMT C4I and leadership. To perform these duties, the Division Chaplain, like the commander, must be positioned wherever the Division Chaplain needs to be to influence religious support coverage throughout the division area of operations. The Division Chaplain, therefore, cannot be tethered to a specific geographic location but must retain the flexibility to respond to changing conditions as they occur in the battlespace. The Division Chaplain must be able to provide C4I while at the same time provide direct religious support to the commander and staff.

5-19. The operational tempo of conflicts will demand 24-hour operations by the units engaged, religious support operations cannot be an exception to this demand if timely religious support is to be provided. Twenty-four hour operations allow for uninterrupted tracking of the battle. The purpose for tracking the battle is to monitor the fight so as to enable timely responses to emerging battlespace conditions requiring religious support.

5-20. In a division fight, 24-hour operations can be confined to a battle-tracking cell if that cell has the common relevant picture and can communicate to outlying UMTs. These outlying UMTs can then maintain 24-hour availability. Battlespace conditions will require 24-hour operations to monitor both current operations and future operational planning.

5-21. Religious support relies upon the availability of sufficient UMT assets to provide coverage. The capability to man 24-hour battle tracking operations cells and the flexibility to reallocate religious support personnel is dependent upon sufficient religious support resources. Flexibility must include the ability to tailor UMTs to match the force packages engaged in the operations.

Chaplain Detachments-Division

5-22. The Chaplain Detachments-Division (CD-D) allows the Division UMT to conduct 24-hour operations. The Chaplain Detachment will come from the Reserve Component and be organic to the Division it supports. This team consists of one chaplain (MAJ) and one chaplain assistant (SSG). The team will drill, train and deploy with the supported division. (See Appendix G and 6-69) Division Chaplains must incorporate the use of CD-Ds into their RSPs and request their deployment through proper staff actions.

5-23. Flexibility must include provisions to transition offense, defense, stability, and support missions.

5-24. Additional planning considerations include—

- The HQ may be structured for incremental deployment.
- Each deployment increment must be synchronized to ensure comprehensive religious support.
- TPFDD information on augmented religious support assets.
- Religious support assets and resource monitoring and management.
- Rules Of Engagement.

Pre-combat

5-25. This phase for the division UMT requires finalized preparation for combat operations. The UMT works closely with the staff, especially the G3/J3, monitoring the tactical situation for last minute changes to the religious support plan. The chaplain assistant NCOIC ensures that—

- Logistical requirements are finalized.
- Battle tracking requirements for the chaplain section are finalized and prepared.
- Appropriate support to Information Operations is coordinated.
- The NCO support channel is advised and monitored for support.
- Personnel issues for the division chaplain assistants are resolved and a contingency replacement plan is staffed with G-1/J-1.

5-26. UMT planning considerations include—

- Spiritual care to subordinate UMTs.
- Spiritual care to division command and staff.
- Managing the implementation of the religious support plan.
- Monitor the current operations situation applying METT-TC.
- Ensuring support for combat operations religious support plan.
- Division chaplain section personnel are spiritually, emotionally, and physically prepared for the operation.

Operations (See Figure 5-1.)

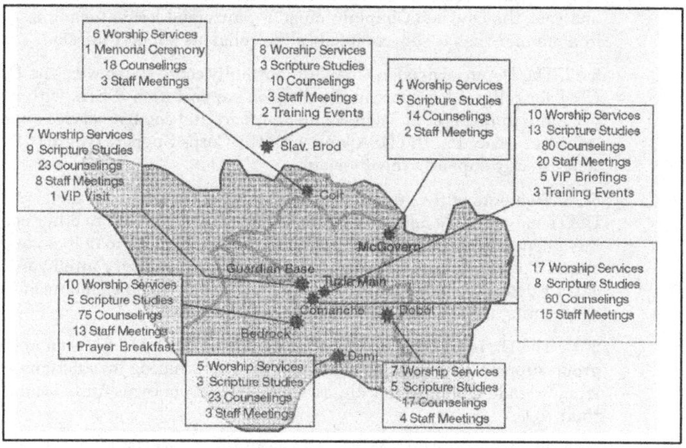

Figure 5-1. Week of Religious Support in Task Force Eagle

5-27. *Critical capabilities requirements.* The ability to provide religious support in the 21st century battlespace will depend upon five critical capabilities requirements: information access, situational awareness, communication connectivity, mobility, and force structure. A deficit in any of these five requirements will radically impair the ability to provide timely religious support.

5-28. Division UMTs must have real time operational awareness. They must be able to determine an appropriate response and expeditiously communicate that information to religious support personnel who will respond directly to the situation. To enable religious support throughout the battlespace, UMTs must be able to move and communicate seamlessly throughout the AO.

5-29. Battle tracking operations require sufficient personnel and the required technical training to conduct 24-hour operations.

Offense

5-30. The Division Chaplain coordinates and supervises religious support for the comprehensive support of all soldiers and units in the division. Based on the religious support estimate, the division chaplain determines area support requirements and where to position the division UMT to best accomplish its mission.

5-31. Normally, the Division Chaplain monitors the tactical situation, plans religious support, and provides care for soldiers from the main Command Post (CP). Prior to an operation, the Division Chaplain moves forward and provides pastoral care to brigade UMTs. Division UMT personnel, under the

Deputy Division Chaplain, provide religious support to soldiers in the rear CP; monitor the volume of casualties, replacements, logistics; and coordinate religious support in the division rear area. These are norms, but in the final analysis, the Division Chaplain must organize and locate his ministry team in a manner that best accomplishes the religious support mission.

5-32. The Deputy Division Chaplain normally coordinates with the DISCOM UMT for religious support in the division support area (DSA); with maneuver units in reserve, division combat support and combat service support units not located in the DSA; and with the Corps Support Group (CSG) UMT that directly supports the division.

5-33. Elements of the supporting CSG may be located in the division AO. Its UMTs may provide additional support to the division and to other corps elements operating in the division area. Other corps units may include the Armored Cavalry Regiment (ACR), artillery, signal, military intelligence, medical, air defense artillery, aviation, and engineers. Some elements of these corps units may be forward in brigade support areas.

5-34. The Division UMT coordinates with the CSG UMT for area and faith group support. The division RSA includes coordinating instructions authorizing brigade UMTs to coordinate with UMTs from corps units located in their AOs.

5-35. The Division UMT coordinates with the Corps UMT concerning area support of corps units in the division AO. The tactical situation may prevent corps UMTs from moving forward to their subordinate units. The teams may be separated from their forward elements by as much as 150 kilometers.

5-36. Area support is limited in light infantry divisions because of restricted transportation assets. Non-divisional UMTs with greater mobility may be tasked to provide more area support to a light division than they would to a heavy division.

Defense

5-37. Division religious support during the defense is similar to religious support during the offense. Access to subordinate UMTs is easier, but enemy reconnaissance and preparatory fires may complicate travel. As in the offense, the division UMT coordinates with the CSG UMT for mutual support.

5-38. The DISCOM UMT coordinates area support in the DSA.

Stability

5-39. Conducting religious support during stability operations remains fundamentally the same as in the offense and defense. The management of the religious support plan is challenged by the increased information management tasks required to properly respond to religious support issues. In stability operations, as in war, the division conducts simultaneous operations. For example, the division could conduct peacekeeping operations at the same time it is deploying additional forces and securing facilities. Stability operations are usually complex. Rarely will UMTs be given a single task to accomplish. Their capability to handle multiple tasks across a wide range will challenge them. When the division headquarters is deployed for stability operations, it is often assigned as the ARFOR and works for a joint task force commander.

5-40. Although often only part of the division deploys, as a controlling headquarters, it may receive attached or operational control (OPCON) units from many external sources. These sources include Army organizations, other U.S. military services, and multinational forces. Within these units are other UMTs that the division UMT must integrate into their religious support plan. Planning considerations for the division UMT—

- ROE understanding and religious support application.
- Command understanding of the religious culture and traditions.
- Media and Public Affairs Office/Officer (PAO) guidance.
- Religious support integration during Base Camp planning.
- Command and staff expansion for operations.
- Information management.
- Personnel issues for UMTs
- Increased liaison tasks with faith-based NGOs and IOs, and host nation religious leaders.

Support

5-41. Support operations consist of relief operations, support to WMD incidents, support to civil law enforcement, and community support. All are challenging for religious support planning and execution. Support operations can also be conducted simultaneously with offense, defense, and stability operations. Posse Comitatus (the legal guidelines defining the relationship between federal armed forces and the civilian populace) plays a critical part for planning. (See Annex K for more information.)

5-42. Religious support planning may include—

- Limits on deployed religious support personnel due to statutory agreements and other restrictions.
- Availability and type of transportation assets.
- Force protection requirements.
- Training requirements.
- SOF participation.
- Use of reserve component forces.
- Other U.S. government agencies' roles.
- Lodgment sites and responsibilities.
- Host nation requirements and expectations (OCONUS).
- Political and social implications of the division's actions.
- Funding and special equipment and supplies requirements.
- Impact of mass casualty upon the units.

Reconstitution, Redeployment, and Demobilization

5-43. Reconstitution is an extraordinary action that commanders plan and implement to restore units to a desired level of combat readiness. It transcends normal daily force sustainment actions. No resources exist solely to perform reconstitution; rather existing systems and units do it.

5-44. Reconstitution decisions belong to commanders. They control assets and decide whether to reorganize or regenerate a unit. Normally, the decision to reconstitute is made two echelons above the unit that must be rebuilt. The commander two echelons above, with his staff's assistance, are in the best position to assess a subordinate unit's effectiveness. The division UMT provides unit ministry support, particularly for cases of combat and operational stress, spiritual renewal, critical event debriefings, and worship opportunities. The chaplain coordinates worship and memorial services, sacramental acts, and pastoral counseling. Chaplains play a key role in assessing the spiritual morale and moral/ethical needs of the unit.

5-45. The OPLAN includes enough details to enable staffs and supporting units to prepare for rapid restoration of units within command priorities. The division UMT inputs the religious support plan in support of reconstitution operations after utilizing the estimation and analysis process. While the religious support plan cannot meet all the contingencies of Army operations, it must be one that the command and UMT can adapt to the situation. In addition to addressing reconstitution for a specific mission, the staff may have to write a separate OPLAN for reconstitution operations. The UMT is fully integrated within this process. The more fully developed a UMT's reconstitution SOP, the easier for it to develop a reconstitution plan.

5-46. Redeployment is the repositioning of forces after mission accomplishment or transfer of mission and authority to another unit. This may mean redeployment to another AO for subsequent missions, or return to home station. In either case, the division UMT should prepare subordinate UMTs to assist commanders, soldiers, and family members, in preparing emotionally and spiritually for the changes that come with redeployment and reunion.

5-47. Demobilization is the return of forces to a peacetime posture. Planning for demobilization is critical. The division UMT cannot wait for the operation to end to begin planning for demobilization. It must begin early in the process. The division UMT's role in demobilization may be much greater than in mobilization. It begins in the division's AO with reconstitution of attached Reserve Component forces and preparation for redeployment to CONUS or OCONUS demobilization stations. Demobilization planning begins concurrently with mobilization. Demobilization planning may impact on division UMTs in theater as reserve component units begin the tasks of reconstitution and demobilization, while under division or corps control. Religious support plans must address many issues including—

- UMT spiritual reconstitution.
- Decorations and awards.
- Evaluation reports.
- Memorial Services and Ceremonies.
- Critical event debriefings.
- Reunion briefings for command and staff, soldiers, and family members.
- The division UMT may also coordinate with the corps and installation chaplain at both the division's home station installation and the RC unit's demobilization station.

Garrison Operations

5-48. During garrison/installation operations, the division UMT plans and implements the CMRP in coordination with the unit training schedule to support the religious needs of the division's soldiers and their families. The division UMT manages and supervises the subordinate UMTs assigned to the division for the commander and coordinates ministry as applicable with the garrison/installation UMT. The UMT participates in all installation chaplains' training programs and other programs that enhance the religious quality of life of the installation. They also fully participate in the unit's training program. The division UMT will support the unit's training program by being available to conduct training for the command and staff, and subordinate UMTs. The division UMT is the mentor for the battalion/squadron UMTs of the division and will execute a mentorship program. Also, the senior division UMT supervises the personnel assigned or attached to the division chaplain's office.

ROLE OF THE DIVISION UNIT MINISTRY TEAM

5-49. The role of the division UMT is critical to the professional and tactical proficiency of the subordinate UMTs. Their leadership responsibilities shape and guide brigade, battalion, and squadron UMTs role in religious support execution process. They are the command and staff advisors for the officers and NCO chain of command.

PRINCIPLES

Table 5-1. Division/Task Force Tactical UMT

CHAPLAIN (O-5) *CHAPLAIN (O-4)*	*CHAPLAIN ASSISTANT (E-8)* *CHAPLAIN ASSISTANT (E-5)* *CHAPLAIN ASSISTANT (E-3)*
Role Commissioned Officer/Staff Religious Professional Title 10 Protected	***Role*** Noncommissioned Officer Specialized Assistant Religious Support Specialist Team Security
Mission Advises commander on religious support and quality of life issues to support the unit's mission. Performs/provides/coordinates the religious support mission for the division. Division Team Leader for Religious Support.	***Mission*** Advises Chaplain and NCO Chain. Provides specialized assistance in areas of religious support to support the chaplain. Perform Battle Staff functions. Team NCOIC.
Parallel Tasks 1. Assess subordinate UMTs readiness. 2. Conducts staff assistance visits. 3. Determines Training Objectives. 4. Mentors subordinate UMTs. 5. Recommend personnel assignments.	***Parallel Tasks*** 1. Assess subordinate UMTs readiness. 2. Conducts staff assistance visits. 3. Conducts UMT religious support training. 4. Mentors subordinate UMTs. 5. Recommends Chaplain Assistant assignments.

5-50. The Division Chaplain is a personal staff officer, religious leader, and team leader. The Division Chaplain Assistant NCOIC must be a NCO leader and advisor, mentor, information manager for the team. They both must have a keen awareness of the world situation as it bears on potential military operations, know the role of military religious leader's and have a broad understanding of the impact of indigenous religions on military operations.

5-51. Unit Ministry Teams must use technology to enhance the personal delivery of religious support. They must be trained in joint and multi-national operations and skilled in synchronizing all aspects of religious support during military operations. They will operate in an environment of increasing volatility, complexity and ambiguity. The division UMT must be innovative and disciplined.

5-52. This means that the Division Chaplain section NCOIC will need to be a graduate of the Battle Staff NCO Course (with the 2S identifier). This will enable the UMT NCOIC to take part in the MDMP in the absence of the chaplain, track the battle analytically and better utilize the digital enablers that the division will share with other staff sections.

5-53. Division UMTs require the same level of technology as the command of the supported force. UMT automation, communication and transportation requirements must keep pace with the command's requirements in order to effectively coordinate religious support, communicate critical information, and remain physically flexible to meet requirements across the division battlespace. They are an integral part of the division.

RESPONSIBILITY AND ACCOUNTABILITY

Staff Responsibilities

5-54. The Division UMT works directly for the commander representing the religious support dimension of caring for soldiers and family members. The UMT executes the commander's plan for religious support initiatives and programs. Staff responsibilities include—

- Synchronizing religious support programs across the division AO.
- Coordinating denominational coverage.
- Performing UMT personnel tasks.
- Advising the Division Commander on matters of religion, ethics, morals, and morale as affected by religion, the impact of indigenous religions on military operations, and installation and community religious support issues.
- Preparing the staff religious support estimate and appendix for the division OPORDs.

Relationship to the Corps Unit Ministry Team

5-55. The Division UMT supports all of the Corps UMT programs including training, family support, and professional development. They work closely together implementing UMT personnel issues for the command. During combat operations, the division UMT utilizes the Corps UMT's religious support appendix to the corps OPORD for religious support planning. The Division UMT works closely with the corps to ensure the success of the religious support mission.

SUPERVISORY RESPONSIBILITIES

5-56. The Division UMT provides technical and professional supervision to subordinate UMTs in the division. As the senior UMT, the Division UMT provides leadership for UMTs in professional development, implementation of the CMRP, personnel issues, and other issues as directed by the commander. The Division UMT works closely with subordinate UMTs to mentor and provide direction.

TRAINING

5-57. The Division UMT assesses its current training strategies as well as the underlying assumptions regarding pre-training knowledge and skills in light of the Division Commander's training and mission requirements. The Division UMT will provide training in planning, preparation and execution of the 11 Religious Support Activities in accordance with the CMRP. Information literacy requirements for all UMTs are imperative in training.

5-58. Training will be profoundly affected since non-linear operations complicate the organization of the battlespace into close, deep and rear operations.

SUMMARY

5-59. Recent significant changes in international politics have created an ambiguous security environment in which threats to world stability and U.S. national interests are more regionally focused. Many crisis situations in which the United States exercises its power span the spectrum of operations from operations other than war to actual combat. Force Projection has become our chief strategy for supporting the power-projection element of U.S. national security. Army corps and divisions must be prepared for force projection operations, as part of a joint or multinational effort and in war as well as other than war environments.

5-60. The mission of the Division UMT is to coordinate religious support to soldiers, families, and as the force structure evolves, to forward-deployed DOD authorized civilians. Division Chaplains also serve as critical advisors to commanders at all levels of the division providing essential information on troop and unit morale, quality of life matters, and the impact of indigenous religion on military operations.

5-61. The Division UMT is expected to operate with a great deal of freedom and responsibility. From information available, it must discern what is essential for the commander's decision making. It must coordinate the entire division religious support effort to support the commander's intent and concept of operation. The UMT develops plans and conducts necessary coordination between subordinate and higher UMTs to execute the commander's intent.

5-62. Combat power in the 21st century will rely heavily upon technology. The cumulative affect of the numerous factors will increase the spiritual, psychological and interpersonal demands placed upon soldiers of the 21st century. Technology will provide additional means to plan, coordinate, and execute religious support to address these demands but the personal delivery of religious support by the UMT will always be imperative.

RESOURCES

5-63. The following is a list of main resources utilized for writing and updating this doctrine.

- JP 1-05.

- JP 5-00.2.
- JP 3-05.
- FM 16-1.
- FM 100-5.
- FM 3-0 (100-5).
- FM 25-100.
- FM 25-101.
- FM 71-100.
- Lessons Learned from past and current operation
- Lessons Learned from the training centers

Chapter 6

CORPS/ECHELONS ABOVE CORPS RELIGIOUS SUPPORT OPERATIONS

The corps is the Army's largest tactical organization. The corps organization is tailored for the assigned theater of operations and the assigned mission. Corps may be assigned divisions of any type required by the theater and the mission. They have a support command and will receive additional Combat Service (CS) and Combat Service Support (CSS) organizations based on the assigned mission.

Echelons above corps (EAC) consist of the Army at theater level and above. These units will be under the command of a regional combatant commander (unified command). This level is in the strategic context.

CONTENTS

OPERATIONAL ECHELON DESCRIPTION

6-1. Corps conducts operations with joint, interagency, or multinational forces, facilitated by information superiority across the full range of military operations, in a Major Theater War or small-scale contingency, against any threat, in any assigned AO. Corps operations may include a mixture of offense, defense, stability, and support operations executed sequentially or simultaneously.

6-2. Corps is usually the link between the operational and tactical levels of war. Corps may link the operational level of war to the strategic level of war. Corps plan and conduct major operations and battles. The corps headquarters synchronizes tactical operations including maneuver; fires, joint, deep, close, and Suppression of Enemy Air Defenses (SEAD); CS; and CSS. The corps is not a fixed standing organization, but a formation whose core characteristic is the ability to form task forces focused on mission requirements. It employs tailored forces that are optimized for specific functions or tasks. The corps employs divisions and applies force allocation rules of support to divisions based on mission. The corps normally has a wide range of Army capabilities to support and sustain any effort. The two critical requirements of corps operations are a robust C4 intelligence, surveillance and reconnaissance (C4ISR) and logistics support to plan, prepare, execute, and assess the full range of operations. Corps traditionally can employ multinational forces and operate with governmental and non-governmental agencies and capabilities. Corps operations are nested in political military dimensions of the combatant commander's theater effort, but are optimized for conducting military operations.

6-3. The corps is a much diversified organization that fulfills the synchronization of many functions that achieve tactical and operational advantage. The corps conducts planning together with joint forces to achieve campaign objectives and conducts simultaneous operations in depth across the corps battlespace. The corps integrates all joint combat, CS, and CSS efforts.

6-4. The corps plans and conducts campaigns and major operations. Corps campaign plans set broad long-term goals, such as control of a geographical area, defeat of an enemy in theater, or the reestablishment of international boundaries. The corps also plans and executes tactical-level battles that support the corps operational objectives, simultaneously conducting close, deep, and rear operations, as well as planning future operations.

6-5. Current U.S. collective security arrangements make it imperative that corps are ready to deploy in response to worldwide situations. In the event of a crisis situation, a corps headquarters considers the employment of light forces, armored maneuver forces, and special operations forces. (See Figure 6-1.)

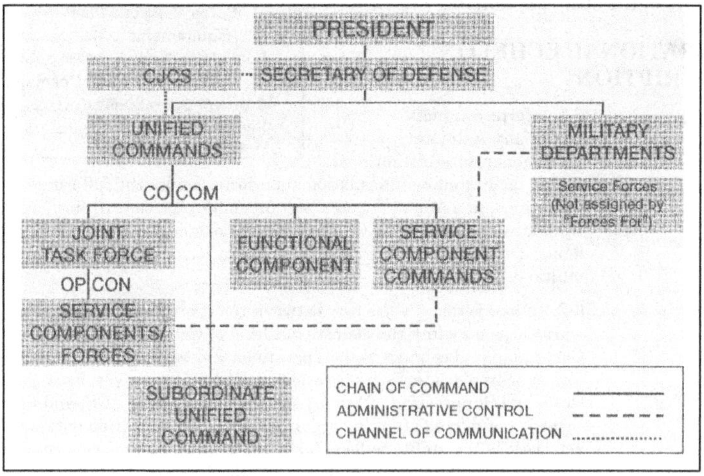

Figure 6-1. Chain of Command and Control

JOINT TASK FORCE CONFIGURATIONS

6-6. A corps headquarters may function as the Army component headquarters of a joint force, the Joint Forces Land Component Command (JFLCC) headquarters of a JTF or as the JTF headquarters itself. The corps' primary responsibility in the joint environment is to participate in the development of supporting joint campaign plans. Current U.S. collective security commitments and the very nature of the international political environment ensure that U.S. corps' will also operate as part of multinational coalition forces.

6-7. Corps with missions to respond to worldwide situations will frequently fight as the largest ground formation in the theater. Often this situation results in the corps commander filling the position of the commander JTF (CJTF). In such cases, the corps headquarters accepts the responsibility to plan operational and tactical level missions as well as oversee the tactical execution of campaigns or other extended operations.

6-8. As the CJTF, the corps commander will exercise OPCON or tactical control (TACON) of assigned forces. The CJTF must form a joint staff and establish the necessary joint boards and centers, as the mission requires. Joint service representation on the staff should be in relative proportion to the service composition of the JTF. It is imperative that all of the staff sections and agencies have joint service representation in the formation and execution of the joint campaign plan. The corps staff must transition to a joint staff structure to ensure unity of effort within the task force.

6-9. The JTF establishing authority is responsible for providing personnel and resources for the corps when the corps is designated the JTF headquarters. However, the corps commander, as the CJTF, must determine what augmentation requirements he needs for the task at hand and coordinate support through the establishing authority. This augmentation is essential in the transition of the corps to a JTF structure.

6-10. Augmentation must be tailored for the specific situation and will be different for every mission. The following areas usually require augmentation:

- Joint, special staff, and personal sections.
- Specific functional area augmentation, such as CA, movement control, and PSYOP.
- Headquarters life support functions.
- Communications support.
- Security support for the JTF headquarters

6-11. The combatant commanders usually organize augmentation for the joint staff into packages based on the nature and tempo of developments and the existing capabilities of the headquarters designated as the JTF. These packages fall into three broad categories:

- Quick response cells to rapidly augment the JTF headquarters with joint and operational area expertise during the early stages of a crisis action situation.
- Functional area augmentation to provide the JTF with expertise not organic to its staff or to enhance specific capabilities based on mission requirements

- Joint staff augmentation to provide the coordinating personal and special staff sections expertise in joint procedures and service-unique capabilities.

JOINT TASK FORCE STAFF AUGMENTATION

6-12. Augmentation facilitates coordination between the JTF staff and the combatant command joint staff, especially with regard to accessing information and capabilities available at the combatant command level.

6-13. Although the corps/ JTF commander applies the military element of power, he also considers those interagency, political, economic, and other resources that may be available and appropriate for the task at hand. The CJTF must understand the strategic and regional environment to include U.S. policy, treaty commitments, the status of forces agreements (SOFA); host nation concerns, religious and cultural considerations, and coalition members' interests. These influences will affect campaign and operational planning, and the establishment of ROE for the joint force. The CJTF must keep in mind the strategic end state through out the planning and execution of the operation.

THE CORPS AS AN ARMY FORCES/JOINT FORCE LAND COMPONENT COMMANDER HEADQUARTERS

6-14. A JTF-level service component command consists generally of forces from only one service of the JTF. The CJTF may designate the corps headquarters as the ARFOR headquarters giving it Title 10 responsibilities and, in most cases, OPCON of a number of assigned or attached forces.

6-15. The designation of Commander, ARFOR carries additional responsibilities for the Corps Commander. The corps as ARFOR must accomplish the same coordination as covered above. Corps headquarters are uniquely suited to be the ARFOR or Joint Force Land Component Commander (JFLCC) headquarters in a JTF. Joint Pub 1-02 defines the Joint Force Land Component Commander as "the commander within a unified command, subordinate unified command, or a JTF responsible to the establishing commander for making recommendations on the proper employment of land forces, planning and coordinating land operations, or accomplishing such operational missions as may be assigned. The joint force land component commander is given the authority necessary to accomplish missions and tasks assigned by the establishing commander. The joint force land component commander will normally be the commander with the preponderance of land forces and the requisite C2 capabilities." The primary purpose of a JFLCC is to provide unity of command for employing land power for the benefit of the joint force as a whole and to give the Joint Force Commander (JFC) a workable span of control.

6-16. The JFLCC must shape the battlespace for subordinate units by focusing on deep operations in space and time (more than 96 hours out).

6-17. Subordinate units are organized for executing the battle command functions required from the higher headquarters of ground maneuver forces. Corps headquarters can coordinate the sustainment activities for significant numbers of assigned or attached forces. Corps also has the staffing to do joint and multinational force coordination and liaison. In most cases the corps headquarters is capable of being the ARFOR or JFLCC headquarters in a JTF without significant personnel augmentation.

THE CORPS AS A FORCE PROVIDER

6-18. In day-to-day operations the corps is required to provide resources to support operations in which the corps headquarters is not otherwise involved. (As an example, each U.S. corps today has soldiers deployed around the globe.) As a force-provider, corps has considerable assets, which they may commit to other commands to support major and lesser regional contingencies around the world. Although these deployed organizations may be under the OPCON or TACON of another headquarters, the corps retains significant responsibilities as the parent organization.

6-19. Operations in stability and support environments are good examples in which corps' function as force providers. Stability operations and support operations often come with vague or ad hoc chains of command, requiring specific support. These missions are different. They require a variety of corps combat, CS, and CSS units to operate in concert with U.S. and foreign civil agencies of government, international, and private organizations.

ECHELONS ABOVE CORPS

6-20. EAC echelons include the theater command. Within this command are various commands that support theater operations. The Army Service Component Command (ASCC) is an Army component in a unified and subordinate unified command. Army forces, through the ASCCs, provide defenses to all joint force components. The ASCC establishes the link between ARFOR and the joint command. It also plans and executes operations in support of the joint campaign, plans and executes support operations to sustain subordinate ARFOR, and provides support to other services in accordance with executive agent responsibilities. The Theater Support Command (TSC) is an EAC-level support organization that is key to operations for CS and CSS execution. The TSC is structured and organized to deploy tailored functional modules. The ARFOR commander is provided key functional expertise and C2 to support joint forces.

FUNDAMENTALS OF RELIGIOUS SUPPORT OPERATIONS

6-21. The corps can rapidly tailor force packages formed from the full range of Army capabilities, both active and reserve. The corps C2 has the organic capabilities for split-based operations required for simultaneous functions of deploying and employing Army forces. The religious support planning and management mission considers force-tailoring UMTs to perform and provide religious support for all types and sizes of forces in all contingencies. Religious support delivery, coordination, synchronization, and synergy are dependent upon the communication capacity in both digital and analog environments to support decisive operations and to improve the synergy of joint and multinational operations. Superior situational understanding is available through information systems that facilitate the execution of decisive, shaping, and sustainment operations across all dimensions of the corps AO.

CORPS RELIGIOUS SUPPORT

6-22. The Corps Chaplain supports the Corps Commander's operational-level responsibilities and roles by engaging in three operational areas:

- Given guidance by the Corps Commander in coordination with G-5, PAO, CA, the Corps Chaplain establishes links with representatives of joint, multinational, interagency, faith-based NGOs and IOs, and religious leaders of the host nation.
- Plans and executes religious support for corps operations. Monitors religious support in major subordinate commands.
- Executes support operations to sustain subordinate Army forces. (See FM 100-7.)

6-23. From the main CP, the Corps UMT plans and synchronizes religious support in the corps area. The Corps Chaplain monitors the tactical situation and moves as far forward as possible to provide staff supervision. The Corps UMT at the rear CP recommends chaplain and chaplain assistant assignments. It coordinates religious support for displaced persons and also plans religious support for reconstitution. It provides an on-call UMT for the Corps Reconstitution Task Force (RTF).

6-24. The Corps UMT normally passes responsibility for religious support in the corps support areas (CSA) to the COSCOM UMT. The Corps UMT must be capable of simultaneous employment and deployment. Its flexible METT-TC planning speeds the process of force tailoring, while its capabilities allow it to provide religious support to the reception, staging, onward movement, and integration (RSOI) at the lodgment base or in an intermediate staging base (ISB) within a JOA. The Corps Chaplain's section will have secure, robust, and survivable communications networks for transmissions of voice, data, and video.

6-25. The senior echelons above division (EAD) Chaplain will tailor the staffing of religious support personnel to match force packages engaged in operations. Religious support relies upon the availability of sufficient religious support assets and the flexibility to reallocate those assets.

6-26. Flexibility must include provisions to transition between combat and other military operations. Planning considerations include—

- Provide interface/liaison between corps and local religious leaders and faith-based NGO/IOs.
- Support the resettlement of refugees and displaced civilians.
- Reconstitute and regenerate combat ineffective units.
- Manage religious support personnel assignments.
- Contractor religious support planning for subordinate UMTs.
- In coordination with the ASCC chaplain, establishes liaison with senior UMTs of other services for religious support to Army authorized personnel, as well as, religious support to service members of other services.
- Conduct specialized religious support assessment of the AO for the commander and staff.
- Coordinate all chaplains' fund activities within the corps and all budget and funding requirements for the corps UMTs.
- Provide spiritual care to subordinate UMTs.

- Coordinate with the ASCC chaplain for Chaplain Detachments (CD) and Special Medical Augmentation Response Team-Pastoral Care (SMART-PC) usage.
- Provide a UMT training program.
- Plan RS 96 hours to 5 days from execution for the corps fight.
- Utilization of "requests for information" (RFI) to respond to FRAGOs and mission RS implied tasks.

6-27. The Corps Chaplain's staff will be resourced to conduct continuous operations. The Corps Chaplains staff will consist of at least four (4) chaplains and five (5) chaplain assistants and will be augmented by the Chaplain Detachment – Corps. Further augmentation may be necessary if serving as the JTF or JFLCC UMT. The Corps UMT will execute sleep plans to minimize the degrading effects of sleep loss. The Chaplain's section must be equipped to minimize the stress imposed by extreme environmental conditions. The Corps UMT will have the capability to exercise religious support management for the commander from any point within the corps extended area of operations.

THE CORPS/EAC TACTICAL/OPERATIONAL/STRATEGIC UMT	
CHAPLAIN (O-6)	***CHAPLAIN ASSISTANT (E-7/8/9)***
Role Commissioned Officer / Staff member Religious Professional Title 10 Protected	***Role*** Noncommissioned Officer Specialized Assistant Religious Support Specialist Team Security
Mission Advises commander on religious support and quality of life issues to support the unit's mission. Performs/provides/coordinates the religious support mission for the command. Senior Team Leader	***Mission*** Advises Chaplain and NCO Chain. Provides specialized assistance in areas of religious support to support the chaplain. Perform Battle Staff functions. Senior Team NCOIC
Parallel Tasks 1. Conducts staff assistance visits. 2. Determines Training Objectives. 3. Mentors subordinate UMTs. 4. Recommends personnel assignments.	***Parallel Tasks*** 1. Conducts staff assistance visits. 2. Conducts religious support training. 3. Mentors subordinate UMTs. 4. Recommends Chaplain Assistant assignments.

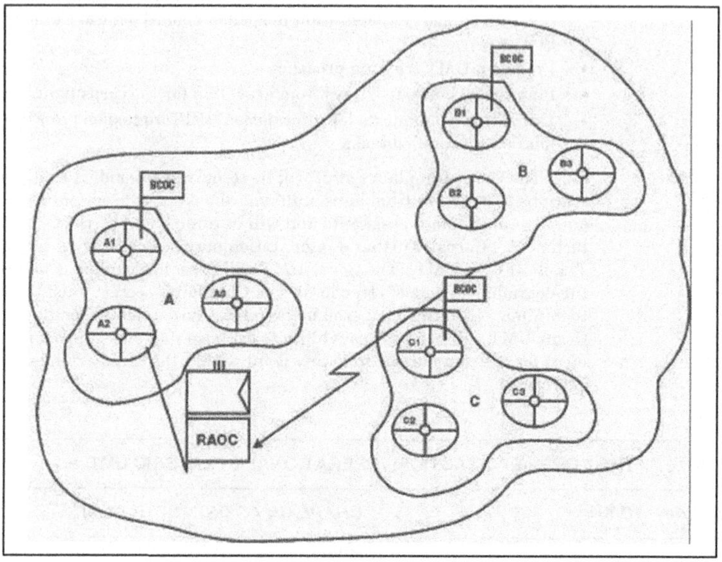

Figure 6-3. Religious Support Coordination in a Corps Support Area

CORPS SUPPORT COMMAND

6-28. The Corps Support Command (COSCOM) UMT coordinates religious support in the corps support area (CSA) for the Corps chaplain. The CSA constitutes the Corps Rear Area. The potential expansion of the CSA, the dispersion of units in bases and base clusters, and the movement of small units make the coordination of religious support difficult. The COSCOM UMT will manage and coordinate area religious support for the Base Cluster Commander to ensure synchronized RS to all units in the base footprint. As a measure of the magnitude of possible expansion, during Operation DESERT SHIELD/DESERT STORM, one COSCOM grew to over 60 UMTs. The area that the COSCOM may cover encompasses the entire battlespace, making authorized and assigned transportation and communication assets for all COSCOM UMTs essential.

6-29. The COSCOM UMT works closely with the Rear Tactical Operation Center (RTOC) and Rear Area Operation Center (RAOC) to track the locations and faith groups of the chaplains in the CSA.

6-30. The COSCOM UMT monitors the flow of religious support supplies with the Corps Material Management Center. The COSCOM UMT locates where communication, battle tracking, and religious support management are best executed. Twenty-four-hour operations are critical to ensure comprehensive religious support and personal staff support to the commander. Split operations at two locations may be considered.

CORPS SUPPORT GROUP

6-31. The COSCOM supports the corps with two or more Corps Support Groups (CSGs): forward and rear. CSGs are composed of multi-functional battalions that provide supply, services, and maintenance to supported units. Forward CSGs support the divisions and other corps units in their AOs. The Rear CSG supports the corps rear area, units in reserve, separate brigades, and units with special missions. The task organization of a CSG can change daily.

6-32. The Forward CSG UMTs monitor the rapidly changing situation through their assigned RAOCs. The tactical communication system found in the RAOC is the primary means of communication for the CSG UMTs. These teams are responsible for coordinating religious support throughout their AOs. They maintain a database with the locations of all UMTs by unit and faith group to process requests for area and faith group religious support.

6-33. The Rear CSG UMT provides religious support to the largest unit in the corps support area that includes the corps and COSCOM headquarters.

6-34. The Rear CSG UMT is often given the reconstitution mission.

RELIGIOUS SUPPORT IN REAR AREAS

6-35. The dispersion of units in rear areas presents a significant religious support challenge. Unit ministry teams have elements of units deployed far forward and out of reach. At the same time, they have elements of other units located in their areas. Of necessity, area support in the rear consumes a greater portion of a UMT's time and energy than does unit support. The Corps Support Chaplain has the responsibility for the rear. Timely religious support coordination is crucial and requires a clear understanding of rear operations.

REAR OPERATIONS

6-36. Rear operations support tactical forces in contact, guaranteeing freedom of maneuver, continuity of support, and uninterrupted C2. In the linear battlespace these actions occur behind forces engaged in active combat. In the non-linear battlespace they occur in the support base areas. Rear operations are characterized by a wide dispersion of multi-functional units conducting a variety of missions around-the-clock.

6-37. Rear areas may be small in a sparse theater, as in small-scale contingencies (SSC), or large in a mature theater, as during general war. In SSC, with tactical operations arrayed through 360 degrees, the "rear area" refers to the area designated as the support base.

6-38. The synchronization of rear operations is the responsibility of a Rear Tactical Operations Center (RTOC), a Rear Area Operations Centers (RAOC), or a Rear Command Post (RCP).

6-39. The Rear Tactical Operations Center (RTOC) is the primary staff element for planning and coordinating rear area security operations.

6-40. The Rear Area Operations Center (RAOC) is a subordinate command post in or near the rear tactical command post concerned with terrain management and security operations.

6-41. The RAOC maintains critical information for the UMT to accomplish its mission: a current situation map; continuous communications with the RCP, other RAOCs, and subordinate base clusters; the threat condition; and Base Defense Status Reports.

> The RAOC maintains critical information for the UMT.

6-42. The most important RAOC function for religious support coordination is terrain management. The RAOC manages the use of terrain by assigning units to bases and base clusters.

Bases

6-43. Units are positioned in bases to share security responsibilities and capitalize on mutual strengths. Bases have clearly defined, defensible perimeters and established access controls.

6-44. Normally, the senior commander is the Base Commander. All forces assigned to the base are under his operational control for defense purposes. The commander forms a Base Defense Operations Center (BDOC) to assist in planning, coordinating, integrating, and controlling base defense efforts.

6-45. The base commander's UMT obtains information about the units (including those in transition) occupying the base from the BDOC. The UMT uses the BDOC's secure communication equipment to coordinate with higher echelons and to coordinate area religious support requirements. (See Figure 6-4.)

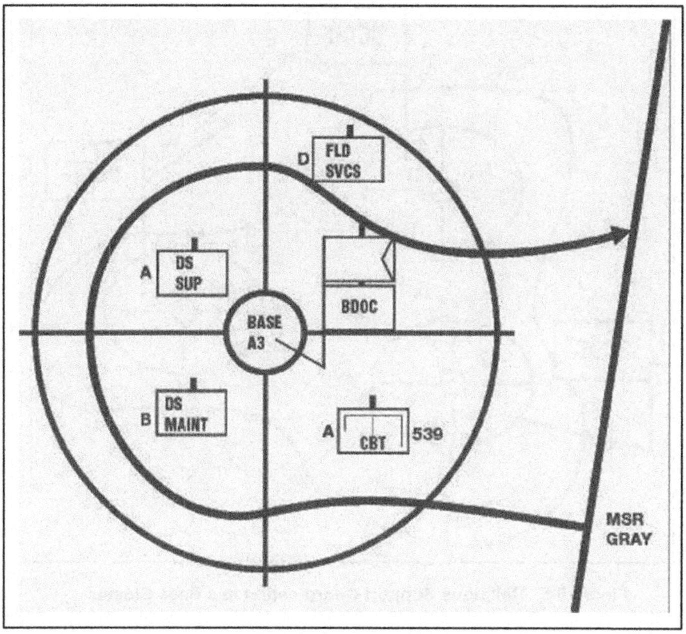

Figure 6-4. Religious Support Coordination in a Base

Base Clusters

6-46. A base cluster is a grouping of bases that are mutually supporting for defense purposes. The commander forms a Base Cluster Operations Center (BCOC) from his own staff and available base assets.

6-47. As in the base, the base cluster commander's UMT uses the BCOC as its source of information for coordinating area religious support in the base cluster and for communicating with higher and lower echelons. (See Figure 6-5.)

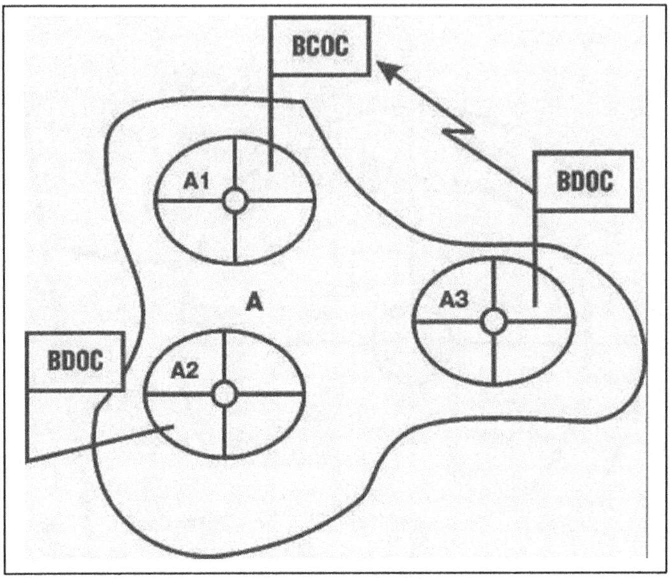

Figure 6-5. Religious Support Coordination in a Base Cluster

Religious Support Coordination In Rear Areas

6-48. The base or base cluster commander's unit ministry team (Base or Base Cluster UMT), regardless of the chaplain's rank, is responsible for area support planning and coordination. Tenant and transient units coordinate with the Base UMT. Base and Base Cluster UMTs coordinate scarce resources for optimum religious support.

6-49. Base and Base Cluster UMTs monitor unit movements, particularly company-size and smaller. They maintain a database of chaplains, their units, faith groups, locations, and means of communication.

6-50. When the commander does not have an assigned UMT, the senior chaplain in the base or base cluster coordinates area support. If there is no religious support available in the base or base cluster, the next higher echelon takes responsibility for area coordination. These relationships are specified in the Religious Support Appendix of the next higher headquarters.

Base Religious Support

6-51. Tenant units coordinate with the Base UMT for additional religious support. The Base UMT conducts a daily meeting to coordinate religious support for the soldiers in the base. Units without organic religious support contact the Base UMT through the BDOC for religious support.

6-52. The Base UMT reports tenant and transient chaplains by faith group to the Base Cluster UMT. This information is used to coordinate faith group support in the base cluster. Unit ministry teams provide area support as needed on a temporary basis until their unit's mission requires their movement from the base.

6-53. If there is no chaplain available in the base to meet a unique faith group requirement, the Base UMT requests support through the BDOC to the BCOC. For instance, if there is no Roman Catholic chaplain available, the Base UMT may request Catholic support using this procedure.

Base Cluster Support

6-54. The base cluster religious support is coordinated in the same way as in the base. The Base Cluster UMT monitors the movement of units into and out of the bases. As bases submit requests for additional faith group support, the Base Cluster UMT meets those requests with available UMTs. If the request cannot be met, it is forwarded to the next higher echelon. The Base Cluster UMT reports chaplains by faith group through the BCOC to the RAOC of the next higher echelon.

Higher Echelon Support

6-55. The procedure for coordination of faith group support described above is repeated up the chain of command. At higher echelons, as the area and the number of units increase, the maintenance of the chaplain database becomes more sophisticated. At these levels, a computerized database, using standard data base software, is required.

SPECIAL RELIGIOUS SUPPORT REQUIREMENTS IN REAR AREAS

RECEPTION, STATIONING, ONWARD MOVEMENT, AND INTEGRATION OPERATIONS

6-56. RSOI operations include the initial reception of units and individuals, the preparation of these units and individuals for combat, and their movement forward into the combat zone. Reception operations begin before hostilities start and continue after hostilities cease. RSOI may be conducted either forward in the theater of operation, in the rear of the area of operation at an Intermediate Staging Base (ISB), or at home station and/or en route to the AO. In the case of RSOI being conducted in the theater, the theater chaplain is responsible for coordinating religious support in the reception area. In the case of a RSOI being conducted closer to the AO, the COSCOM chaplain may be designated to provide area religious support.

MORTUARY AFFAIRS

6-57. The recovery, preparation, and transportation of the dead are one of the most difficult missions in the battlespace. Religious support is provided to soldiers in Mortuary Affairs units and to those detailed to perform this mission. The care for the dead requires sensitive religious support.

6-58. Planning Considerations:

- Religious support in sites where those killed in action are prepared for return to CONUS.

- Religious support for soldiers who are doing the work.
- Special faith group requirements for burial.

RECONSTITUTION AND REDEPLOYMENT

6-59. Reconstitution (both reorganization and regeneration) restores combat ineffective units to a specified level of effectiveness. The Army Service Component Command (ASCC) plans and conducts operational and tactical reconstitution operations.

6-60. Regeneration involves large-scale replacement of personnel and equipment and the rebuilding of units. This mission is usually assigned to the TSC, COSCOM, or DISCOM, depending on the size of the unit. Division, Corps, or ASCC UMTs assist in religious support to the units.

6-61. One or more UMTs may be assigned to a RTF to provide religious support to soldiers until the unit's assigned UMT has recuperated or been replaced. Religious support during reconstitution includes the ministry of presence, group and individual counseling and debriefing (see Appendix C, Critical Intervention Management and Critical Event Debriefings), worship and memorial services, and prayer. RTF UMTs play an important role in restoring the soldiers' spiritual fitness and rebuilding the unit's cohesion and morale.

6-62. Planning Considerations:

- RTF planning process and status.
- How many UMTs are required?
- Which UMTs are designated to join the RTF, how are they notified, and by whom?
- Anticipated duration of mission.
- Criteria for determining when RTF UMTs end the mission and return to their parent units.
- Logistical support for RTF UMTs

THEATER RELIGIOUS SUPPORT

6-63. The Army Service Component Command (ASCC) serves as the senior army echelon in a theater and is the service component command of a unified command. It includes the service component commander and all Army personnel, organizations, units, and installations. The ASCC chaplain is the senior Army chaplain in a theater of operations and provides staff supervision over all Army religious support in the theater. The ASCC chaplain is responsible for recommending religious support policy to the ASCC commander. An example of an ASCC is Third U.S. Army, which is the ASCC of USCENTCOM.

OPERATIONS

6-64. The ability to provide religious support in the 21st century battlespace will depend upon five critical capabilities requirements: information access, situational awareness, communication, connectivity, mobility, and force structure. A deficit in any of these five requirements will radically impair the ability to provide timely religious support.

6-65. EAD UMTs must have real time operational awareness. They must be able to determine an appropriate response and expeditiously communicate that information to religious support personnel who will respond directly to the situation. To enable religious support throughout the battlespace, UMTs must be able to move and communicate seamlessly throughout the AO.

6-66. The complexity of coordinating religious support at EAD requires sufficient religious support personnel to provide flexible tailoring and augmentation to support the mission. The senior theater UMT estimates and plans campaigns with the staff and must focus on the different phases, determine priorities, and manage the coordination and execution of religious support tasks. When the theater commander establishes the combat zone (CZ) and communications zone (COMMZ), the theater UMT plans for simultaneous

6-66. activities across the full spectrum of operations throughout both AOs. Communication across both areas is critical for religious support synchronization; reach capabilities, reporting, and comprehensive planning. Division rear operations and Corps MSR operations are linked as a part of the battlespace management system, which includes tracking religious support assets from COMMZ through assembly areas (AA) to corps rear to division.

STRATEGIC RELIGIOUS SUPPORT PLANNING

6-67. The combatant commander's UMT's strategic religious support planning provides the framework for the religious support assets in peacetime and in crisis. During predeployment, the senior command UMT develops an assessment that transitions the focus to crisis or war. Force projection and assessment transitions religious support assets to combat operations. Religious support planning is integrated into the combatant commander's staff plans and estimation process. The result is expressed in terms of religious support to mission objectives, concepts, vision, resources, and provides guidance to subordinate command UMTs for a broad range of religious support activities. Planning should consider—

- The command's desired end state and how religious support can support full accomplishment of military objectives.
- Risks associated with METT-TC and the religious support assets.
- Religious support courses of action meeting mission objectives.
- Visualization of the religious support needs.

STRATEGIC RS ESTIMATE

- Assigned objectives from combatant commanders.
- Translation of combatant commanders objectives applicable to theater RS.
- Visualization of the strategic environment; how it relates to the accomplishment of the commander's RS objectives.
- Assessment of the threats of accomplishment of assigned RS objectives.
- Assessment of strategic alternatives to available, with accompanying analysis, risks, and the requirements for RS plans.
- Impact of indigenous religions on operations.
- Considerations of available RS resources, linked to accomplishment of assigned RS objectives.

ARMY SERVICE COMPONENT COMMAND RELIGIOUS SUPPORT

6-68. The ASCC chaplain manages UMT, chaplain and chaplain assistant, assignments throughout the theater for the commander. The senior chaplain assistant NCOIC advises the senior chaplain on all chaplain assistant assignments. The ASCC command UMT provides training, policy, logistics, and funding for all Army UMTs in the theater and serves as primary liaison with the next higher commander's staff chaplain. The ASCC UMT stays in close communication with senior UMTs in each sector. They participate in joint planning to ensure that the interest of the commander's religious support goals and plan are represented and implemented. They coordinate with the TSC UMT for movement, and transportation unit information and resources. The Personnel Command (PERSCOM) UMT in theater coordinates assignments for UMTs for the TSC command UMT. Functions of the ASCC UMT include—

- Establish liaison with host nations religious leaders and other Joint Religious Ministry Support Teams at the highest levels in the theater.
- Coordinate fund transitions for logistical support.
- Coordinate religious support for force arrival in transfer of authority (TOA) and RSOI sites.
- Coordinate religious support for contractors for the command as part of the commander's policy.
- Provide guidance to all senior Army UMTs to support the combatant commander's RSP.
- Plan and provide training for UMTs.
- Manage UMT assignments within the theater to ensure appropriate denominational balance to all areas of the theater.
- Provide guidance for the employment of all Chaplain Detachments (CDs) and SMART-PC in support of the religious support mission.

CHAPLAIN DETACHMENTS

6-69. When requested from a MACOM, the HQDA G3 will task USARC through FORSCOM to mobilize and deploy the requested chaplain detachments needed to support planned, current, or ongoing operations. The Chief of Chaplains may request that HQDA G3 task USARC to mobilize and deploy chaplain detachments in support of specific missions. (See Appendix G.)

ADDITIONAL CORPS/ECHELONS ABOVE CORPS RELIGIOUS PLAN REQUIREMENTS

SPECIAL OPERATIONS

6-70. The mission, duration and scope of operations, security considerations, signal capabilities, and the desired degree of control determine Special Operations Force (SOF) command relationships. SOF units above group or regimental level are Joint. Joint force commanders include unified, subordinate unified, JTF, functional component, and commanders of specific operational forces. USSOCOM organizes, equips, trains, and provides Special Operations Force for combatant commanders of regionally oriented unified commands (European, Atlantic, Southern, Pacific, and Central) and for the Commanding General, U.S. Forces, Korea. Combatant commanders delegate operational control of these SOF forces to their theater-level Special Operations Commands (SOCOMs). The six SOCOMs are the focal point for in-theater SOF, form nuclei for the Joint Special Operations Task Force (JSOTF), and furnish expertise needed to effectively employ SOF independently or in concert with conventional forces.

6-71. A unified command normally consists of the four service-specific conventional forces and a SOCOM. The combatant commander or the SOCOM may establish a JSOTF commanded by the Joint Force Special Operations Component Commander (JFSOCC). SOF units from the three services are normally under the OPCON of the JFSOCC.

6-72. ARSOF missions are often highly classified and sensitive. To provide counseling, debriefing, and other religious support to ARSOF units, the chaplain must have a Top Secret clearance. Chaplain assistants are required to have a clearance equal to mission classification. The USASOC UMT recommends all ARSOF UMT assignments.

6-73. Normally, the JSOTF chaplain is a member of the service having the majority of forces. (See JP 1-05, Chapter 2.) If the Army supplies the majority of forces, an ARSOF chaplain is normally designated as the JSOTF chaplain.

6-74. The JSOTF chaplain coordinates with the senior chaplain of the next higher headquarters to ensure complete religious support for SOF units. The JSOTF chaplain, in turn, supervises SOF religious support throughout the theater. The chaplain coordinates for additional religious support when other SOF chaplains are unable to provide religious support. SOF chaplains may provide limited area support for conventional forces when it does not conflict with SOF mission requirements

MEDICAL COMMAND SPECIAL MEDICAL AUGMENTATION RESPONSE TEAMS-PASTORAL CARE

6-75. The U.S. Army Surgeon General (DASG) and the U.S. Army Chief of Chaplains (DACH), in consultation with the Chief of Staff of the Army, the Director of Military Support, and applicable Federal and Department of Defense regulations, requires designated DASG/MEDCOM/DACH assigned, attached or associated individuals and organizations to develop and maintain special organized, trained and equipped SMART-PC.

6-76. Operations consist of—

- Alert, assemble, and issue warning order within 6 hours of no-notice notification.
- Deploy from home station within 6 hours of the warning order.
- Conduct mission: supporting commander without organic ministry team or needing additional ministry team support for mass casualty or devastating contingency operation; or provide professional pastoral care augmentation (to include Critical Event Debriefing, Trauma Ministry, Mass Casualty Ministry, and Spiritual Assessment) to local authorities in the management of disaster/mass casualty incidents.

6-77. SMART-PCs deploy with enough food, clothing, personal hygiene supplies, and equipment for self-sustainment in an austere environment for 72 hours. Team missions normally conclude in 72 hours. When extenuating or mitigating circumstances require DASG/MEDCOM or DACH approval for mission extension, SMART-PC members will procure locally or formally obtain logistical support from supported units, or local civil or Lead Federal Agency. SMART-PC members deploy on orders articulated on DD Form 1610.

6-78. At the direction of DACH and/or on order of DASG/MEDCOM, at request of commanders, and/or the request of legitimate civil, Federal, or Defense authorities, using appropriate, recognized, and approved channels designated units will deploy SMART-PC to provide short duration, medical ministry augmentation to:

- Regional, domestic, Federal and Defense agencies responding to disaster, civil-military cooperative action, humanitarian and emergency incidents.
- Support units without organic ministry teams or who require additional ministry team support for mass casualty or devastating contingency operation.

6-79. SMART-PC is an addition to the list of required teams, which includes:

- Trauma/Critical Care (SMART-TCC).
- Chemical/Biological (SMART-CB).
- Stress Management (SMART-SM). (This team has one CH assigned, no assistant.)
- Medical Command, Control, Communications, Telemedicine (SMART-MC3T).
- Preventive Medicine/Disease Surveillance (SMART-PM).
- Burn (SMART-B).
- Veterinary (SMART-V).
- Health Systems Assessment and Assistance (SMART-HS).

6-80. SMART-PC is modeled on the standard Unit Ministry Team of one chaplain and one chaplain assistant. Operationally, multiple teams are task organized based upon METT-TC and medical-ministry/risk analysis to provide the appropriate level of response and technical augmentation to meet mission needs.

MANAGEMENT:

DASG, ASSISTANT SURGEON GENERAL, FORCE PROJECTION
Policy, supervision, and oversight to equip, deploy and redeploy SMART-PC. Validate and resource special equipment requirements; periodically test and evaluate individual SMART-PC teams; develop, monitor, and report readiness reporting requirements. Establish and maintain officer liaison with DACH.

DASG / MEDCOM, ASSIGNED REGIONAL MEDICAL COMMANDS
Organize, train, and equip 2-SMART-PC teams. On order, deploy assigned SMART-PC within assigned regional medical boundaries. Be prepared to deploy outside of regional boundaries in support of other SMART-PC or to other national/international sites.

EQUIPMENT: Configuration of SMART-PC deployment sets, kits, and outfits will be man portable, and utilize commercial backpacks, suitcases, and rugged containers. Each equipment piece will meet commercial airframe size, weight, and cube requirements.

MEDICAL COMMAND SPECIAL MEDICAL AUGMENTATION RESPONSE TEAM-STRESS MANAGEMENT

6-81. The role of the chaplain assigned to a SMART-SM is to bring spiritual aspects of stress management, as a clinician, in support of the team's overall tasks, which include:

- Critical event debriefings.
- Spiritual counseling.

CONTINGENCY FORCE POOL (CFP)

6-82. The FORSCOM Chaplain's Office will coordinate with CONUSA Chaplains where CFP chaplains are attached for duty to ensure that chaplain personnel are integrated into MSCA contingency training and planning scenarios, and are included in contingency plans. The FORSCOM Chaplain's office will coordinate with the CONUSA Chaplains to de-conflict training and planning objectives. FORSCOM will ensure that CFP personnel support is placed in the Regional Planning Agents (RPA) MSCA handbook, in the DOD database to the Defense Coordinating Office (DCO) and integrate CFP training / certification with DOD Emergency Preparedness Training and CONUSA MSCA contingency exercises. Upon request, FORSCOM has the ability to cut orders for CFP personnel using ARPERCOM Orders and Resource System (AORS) and can make one or more CFP personnel immediately available depending on the crisis level. In a small-scale contingency operation (SSCO) OCONUS response, CFP personnel will report to the CONUS MACOM for contingency support, providing religious support and liaison / coordination. During MSCA or Homeland Defense / Security (HLD / HLS) CONUS contingencies, upon request by a DCO and with appropriate funding for RS to DOD personnel, CFP personnel become assets to the JTF Chaplain.

ROLE OF THE CORPS/ECHELON ABOVE CORPS UNIT MINISTRY TEAMS

6-83. The personal delivery of religious support by the UMT will be imperative in all operations. It is essential that the Corps/EAC HQs UMT is able to rapidly tailor and modularize UMTs for operations. Corps/EAC UMTs must organize to optimize and use information. Flexibility in applying METT-TC is critical to accommodate specific missions. Rank structure and composition of UMTs may change to support specific missions. RS planning must be robust and flexible enough to provide religious support across the full spectrum of military operations. Corps/EAC UMTs require the same level of technology as the supported force. As technology advances the UMT automation, communication and transportation requirements must keep pace with the command's overall requirements.

GARRISON OPERATIONS

6-84. During garrison operations, all EAC UMTs, the Corps UMT, and subordinate UMTs support the training program of the installation chaplain. They implement their CMRP to support the religious needs of the families and soldiers. The EAD UMTs manages and supervises subordinate UMTs assigned for the commander. They support the commander's training program through participation and planning and implementing UMT training. The Corps UMT is a mentor to the BDE UMTs and will execute a mentorship program. The Corps UMT supports the ASCC UMT RS plan.

SUMMARY

6-85. Corps and EAC UMT role is crucial to the warfighting religious support mission success. AC and RC are fully integrated to supply combat ready UMTs to provide religious support to the full spectrum force that meets the needs of the joint force commander in war, conflict and peace. Command UMTs are fully trained, ready, and staff integrated to meet the religious support needs of soldiers, family members, and authorized civilians in all operations. The UMTs are strategically and operationally responsive and prepared to conduct prompt and sustained religious support operations as part of joint, multinational, and interagency teams.

RESOURCES

6-86. The following is a list of main resources utilized for writing and updating this doctrine.

- AR 165-1, Chaplain Activities in the U.S. Army
- JP 1-05, Religious Ministry Support for Joint Operations
- JP 5-00.2, Joint Task Force Planning Guidance and Procedures
- JP 3-05, Doctrine for Joint Special Operations
- JP 3-0, Doctrine for Joint Operations
- JP 3-07.3, Joint Tactics, Techniques, and Procedures for Peace Operations
- JP 1-0, Doctrine for Personnel Support to Joint Operations
- JP 3-08, Interagency Coordination During Joint Operations

- JP 3-10, Joint Doctrine for Rear Area Operations
- FM 1 (100-1), The Army
- FM 16-1, May 1995, Religious Support
- FM 3-0 (100-5), Operations
- FM 25-100, Training the Force
- FM 25-101, Battle Focused Training
- FM 100-15, 29 OCT 1996, Corps Operations
- FM 100-11, 15 JAN 1998, Force Integration
- FM 100-9, 13 JAN 1992, Reconstitution
- FM 100-8, 24 NOV 1997, The Army In Multinational Operations
- FM 100-7, 31 MAY 1995, Decisive Force: The Army in Theater Operations
- FM 100-6, 27 AUG 1996, Information Operations
- FM 22-100, Army Leadership
- DA Pam 165-3, Chaplaincy Training Strategy
- National Military Strategy
- Joint Vision 2010
- Joint Vision 2020
- Army Vision 2010
- Concept for Future Joint Operations, May 1997
- U.S. Army Chaplaincy Strategic Plan FY 2000-FY 2005
- Chaplain Support Force XXI White Paper
- Lessons Learned from past and current operation
- Lessons Learned from the training centers

Chapter 7

INSTALLATION UNIT MINISTRY TEAM RELIGIOUS SUPPORT OPERATIONS

7-1. The Army, Active and Reserve components, is designed, equipped, and trained for force projection operations. Army installations function as force projection platforms, providing support for deployed tactical units throughout all stages of force projection. To ensure readiness, the Army continually plans and trains for the full range of military operations.

CONTENTS

7-2. Combat readiness also produces an Army whose organization, skills, leadership, and discipline can advance U.S. interests in a variety of operations not characterized as war. These operations include, but are not limited to, nation assistance; humanitarian and disaster relief; counter drug operations; peace operations; antiterrorism; shows of force; attacks and raids; noncombatant evacuation operations; insurgency and counterinsurgency support; civil disturbance and support to domestic authorities.

7-3. Force projection is the military component of power projection and is a critical component of U.S. strategy. AC and RC units, the mobilization base, DA civilians, and the industry are directly involved projecting the force anywhere in the world. Training, mobilization, deployment, employment, sustainment, and redeployment are the processes utilized for force projection. They are continuous, overlapping, and interactive throughout an operation. The process of force projection is an integral part of the Joint Operation Planning and Execution System (JOPES). JOPES is used to monitor, plan, and execute mobilization, deployment, employment, sustainment, and redeployment activities associated with joint operations. The Army force projection process is imbedded within this system.

7-4. Religious support is a vital part of the combat readiness mission of the installation in force projection. Force projection operations are inherently joint and religious support planning and synchronization are required.

INSTALLATION DESCRIPTION

MISSION

7-5. Army installations, which include STARCs and Reserve Support Command (RSCs), contribute to the Army's ability to support the National Military Strategy (NMS). Peacetime missions include: counter-drug operations on our national borders, nation building, counter-terrorism, WMD support, deployment of land forces for a conflict, or fighting and winning the nation's war. Each installation, STARC, and RSC has a primary mission that determines its assignment to a MACOM or Component.

ORGANIZATION

7-6. Installations may host dozens of organizations that are either part of the MACOM's mission or simply tenants assigned to the installation for support. Installations are organizationally characterized by the nature of the major activity they host. For force basing they would have Corps and Division(s). For unit training they would host such units, as Combat Training Centers (CTC), for example, the Joint Readiness Training Center (JRTC). For the logistic support they would provide depot and commodity management. Within the Army, an installation may be referred to as a post, camp, station, fort, barracks, subpost, depot, arsenal, proving ground, base, laboratory, port, or ammunition plant. For the Army National Guard and Reserve Component the term "installation" refers to the RSC for the RC and the STARC for the ARNG.

As Part Of A Major Army Command

7-7. Army installations are assigned to and operate under commanders of Major Army Commands (MACOMs) or MSCs where applicable, such as in the Army Materiel Command (AMC), and the U.S. Army Reserve Command (USARC). STARCs operate under the guidance of the National Guard Bureau. Installations consume, provide, maintain, and otherwise control significant resources while executing DOD, DA, and MACOM mandated policy programs. These programs are reviewed, analyzed, and resourced at DA level. Transformation Installation Management (TIM) will eventually transform the relationships between installations and MACOMs, to include, the relationships between MACOM Chaplains and Installation Chaplains. UMTs should keep abreast of such changes.

As Part Of Full Spectrum Operations

7-8. The entire installation management system is a key component of developing and sustaining a total force capable of meeting the full spectrum of missions required by the National Military Strategy (NMS). The Army's mission to expand and rapidly deploy trained and ready units is dependent upon installation capabilities. Within the installation support system is the Army's mobility infrastructure. Total force readiness is linked to installation operations.

KEY TERMS

- Joint Operation Planning Execution System: (JOPES)
- Area Support Group: (ASG)
- Port of Embarkation: (POE)
- Arial Port of Embarkation: (APOE)
- Seaport of Embarkation: (SPOE)
- Port of Debarkation: (POD)
- Aerial Port of Debarkation: (APOD)
- Seaport of Debarkation: (SPOD)

7-9. Army installations are standardized to ensure that operations can be effectively supported in peacetime. This effectiveness must continue throughout the transition from peacetime to wartime (mobilization) and during post conflict (post-deployment). Army installations provide the sustaining support necessary for units to accomplish their missions.

7-10. Installations provide this support by accomplishing both their primary and support missions. The primary mission of the installation includes, but is not limited to, readiness and preparation for operational deployment, training, supply, and/or maintenance. The support mission includes real property acquisition and upkeep, community service, and other personnel support.

7-11. Sustainment includes the primary and support missions of the installation. Throughout each phase of deployment, the installation manages the sustainment of units.

FUNDAMENTALS OF INSTALLATION RELIGIOUS SUPPORT OPERATIONS

7-12. Military chaplains and chaplain assistants are specially trained and assigned to provide for the government's interest to protect free exercise of religion rights in the military, which is a unique pluralistic and multi-faith environment. Military chaplains and chaplain assistants assigned in Table of Distribution and Allowances (TDA) positions provide direct support to the military commanders for the installation religious program, Family Life Centers, units, hospitals, initial entry training (IET) units, Chaplaincy resources management, Chaplaincy personnel management, Chaplaincy force structure and training. The content of the religious support mission of the chaplain and chaplain assistant is no different in the TDA than the TOE; only the command context changes. It must be understood that the Chaplaincy is not a social service under the city management context of the Installation, but a direct support function to the commander. Religious support is the military commander's mission. TDA positions are required to develop military religious support leaders and train religious support skills for the full spectrum of military operations, from the sustaining base to the forward edge of the battlespace.

7-13. Installation religious support has three major areas:

- To prepare and assist soldiers and families for military operations.
- To provide religious support for units and personnel on the installation.
- To sustain religious support operations during deployments and mobilizations.
- To train UMTs for comprehensive RS.

7-14. The installation chaplain is responsible for all religious support on the installation including all religious services and activities. The installation chaplain coordinates with tenant units to implement the CMRP. The installation's UMTs provide a broad religious support programs for units, soldiers, and their families. Worship services, pastoral care, religious education, and spiritual fitness training are provided for the religious support needs of the installation population.

READINESS

7-15. Military chaplains and chaplain assistants in the TDA are necessary to support peacetime deployments, as well as, to man operational units in mobilization and provide wartime rotation, and rapid casualty replacement. All TDA military chaplain and chaplain assistant positions are required to be trained and ready to use combat skills and training to implement the military religious support mission of the commander in any contingency.

7-16. The Reserve Component (RC) holds a significant portion of the mobilization assets for the Chaplaincy. The RC Chaplaincy is a military combat augmentation in direct support of the commander. RC chaplains and chaplain assistants in the TDA are necessary to support peacetime deployments, as well as, to man operational units in mobilization, provide wartime rotation and rapid casualty replacement. All RC TDA chaplains and chaplain assistants are required to be trained and ready to use combat skills and training to implement the religious support mission of the commander in any contingency. The RC Chaplaincy provides the recruitment base for both the AC and RC. The RC provides the training base for the Chaplain Candidate Program. The IMA and Active Guard Reserve (AGR) chaplains and chaplain assistants are in the TDA structure.

Assessment of Unit Ministry Team Readiness

7-17. As the commander's personal staff officer for religious support, the command installation chaplain is responsible for assessing the technical proficiency of UMTs in subordinate and tenant units. They conduct assessments utilizing the units training program such as during field training exercises (FTXs), external Army Training and Evaluation Program (ARTEP) events, unit training, Reserve component annual training and Command Inspections. Observer controllers may provide additional assessment assistance when the UMT is involved in Battle Command Training Program (BCTP) events or Combat Training Center (CTC) rotations. They acquire, in coordination with the senior unit UMT lessons learned to integrate into the installation UMT training program. (See DA PAM 165-3, The Chaplain Training Strategy.)

7-18. Chaplains and chaplain assistants participate in the installation's Command Master Religious Plan (CMRP) to help soldiers develop the personal spiritual readiness needed to sustain them during combat. Spiritual readiness is also essential to sustain soldier families while the unit is deployed. As part of the installation's CMRP, the UMTs conduct worship services, offer religious education, and provide spiritual fitness training and other activities in installation chapels and facilities.

Training

7-19. Training is the cornerstone of preparation for military operations. The Army's primary activity during peacetime is realistic, battle-focused training oriented on the unit's mission essential task list (METL). Installation chaplains are responsible for individual training, MOS 56A and 56M. The installation chaplain's training managers coordinate with subordinate UMTs and unit S3/G3s and ensure that UMTs have adequate technical and tactical training for deployment and combat. This is implemented through the development and execution of an annual Installation wide UMT training strategy and plan (the religious support METL provides training tasks along with the Chaplain Officer and Enlisted Critical Task Lists). This training is integrated within the implementation of the Chief of Chaplain Training Strategy (DA Pam 165-3) and higher headquarters training strategy.

7-20. Tactical proficiency is a demonstrated understanding of the Army's war-fighting doctrine and tactics. To provide effective religious support in combat, the UMT must know its unit's doctrine, tactics, techniques, and procedures. The UMT must also maintain the requisite skills of field craft essential for survival in the battlespace. Tactical training is primarily the responsibility of the unit to which the UMT is assigned or attached.

7-21. Technical proficiency is the demonstrated ability to perform all tasks related to one's functional area. Technical proficiency is required for both garrison and field missions. Technical proficiency is acquired from schools, self-study and training with other UMTs in performing the unit's and installation's religious support mission. Training and continuing education conferences, functional courses offered by USACHCS, and courses offered through the AMEDD Center and School, and the NCO Education System (NCOES) supplement the installation religious support training program.

7-22. UMTs must balance religious support training for combat and for garrison. The UMT must participate fully in both unit and religious support training to assure their tactical and technical proficiency. If training for either combat or garrison operations is neglected, ministry teams will not be prepared to provide comprehensive religious support to soldiers and their families.

7-23. Many installations serve as mobilization bases. (See Figure 7-1.) Mobilization can overwhelm the installation's resources. Thorough planning and realistic training reduces confusion. Mobilization training gives UMTs an opportunity to practice their mobilization religious support missions and an opportunity to assess the adequacy of mobilization plans. Installations assist with the following RC UMT training events:

- Annual training (AT).
- The Individual Mobilization Augmentee (IMA) training.
- Individual training events, Individual Ready Reserve (IRR), and chaplain candidate training.
- AC/RC Sustainment Training.
- Annual Chaplaincy Regional Sustainment Training (CREST) as mandated by the Chief of Chaplains.

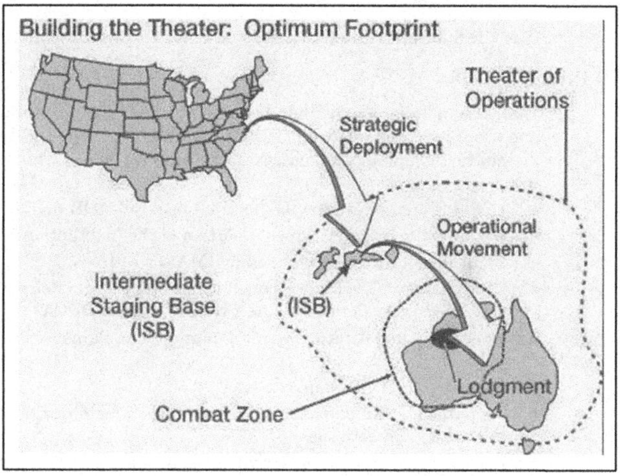

Figure 7-1. Building the Theater

FORCE PROJECTION

7-24. When the installation is serving as a force projection platform, the installation UMT must be prepared to provide as a staff the religious support mission. The installation chaplain's staff must be prepared to support all phases and sustainment requirements of the mission. Provision is made for religious support for all phases of deployment including pre-deployment, movement to the POE, strategic lift, theater reception, and theater onward movement. This includes counseling, training of UMTs, leadership training, reunion briefings and counseling services (and as a family readiness group resource). Installation UMTs will support peacetime deployments, as well as, to man operational units in mobilization, provide wartime rotation and rapid casualty replacement.

7-25. Force Projection planning includes—

- Mobilization, receiving, housing, processing, validating, deploying of soldiers including UMTs.
- Support the increased information communication requirements that will occur.
- Assist in reception as the force shifts MRC priority and re-deployment of forces.
- Maintain installation religious support requirements for the remaining military community after deployment.
- Training of UMTs as required.

MOBILIZATION PHASES

(See CH 3 for more information.)

PLANNING PHASE

7-26. The installation chaplain becomes the mobilization station chaplain and assumes a key role when RC UMTs mobilize. Before mobilization, the installation chaplain designates a mobilization planner who does the following:

- Updates and maintains the installation's RC MOB Plan book.
- Updates the religious support portion of the installation MOB PLAN.
- Identifies positions to be filled by IMAs or retirees.
- Clarifies cross-leveling responsibilities and coordinates with the G1/Director for Personnel and Civilian Affairs (DPCA).
- Develops a mobilization-training plan for chaplains and chaplain assistants.
- Prepares UMTs for deployment.
- Prepares for and participates in installation mobilization exercises (MOBEX) and activities.
- Plans for demobilization activities such as after action reviews, reunion briefings, awards, reports, etc.
- Maintains and improves combat readiness posture.
- Ensures all sites, such as RSOI, are fully manned by UMTs during all phases of mobilization, deployment, and redeployment.
- Family Readiness Group (FRG) religious support.
- Prepare the UMT MOBTDA.
- Coordinates with and tracks RC units with and without UMTs scheduled to report to the installation.
- Communicates with IMAs and retirees with orders to report to the installation upon mobilization.

7-27. This phase concerns all AC and RC UMTs efforts at all levels during peacetime to plan and train for assigned wartime tasks. The installation command UMT should ensure that the installation RS mobilization plan addresses civilian support for the mission. They must review the plan annually ensuring it is up-to-date and fully integrated into the installation commander's mobilization plan.

7-28. STARC UMT mobilization plans and SOPs clarify training and MOBTDA personnel to provide religious support to families at family assistance centers (FACs). Important consideration must be given to identifying the religious support requirements of the FACs and developing MOBTDAs with sufficient UMT personnel support.

7-29. The CONUSA UMT coordinates between the FORSCOM staff UMT and RC UMTs within its geographic area. CONUSA staff chaplains and NCOICs may redistribute UMT members in order to match anticipated UMT mobilization requirements.

ALERT PHASE

7-30. This phase begins when an AC or RC unit receives notice of a pending mobilization order. During this phase, the STARC and Army Reserve Command chaplains and NCOICs continue to cross-level individual members of UMTs within their state or region to bring alerted units to required religious support deployable status.

HOME STATION PHASE

7-31. This phase begins with the RC unit's entry on active federal duty or the AC's preparation for deployment. Actions are preformed to bring units to active status. Installation, STARC, and USARC Command UMTs continue cross-leveling, as well as, ensuring critical UMT logistical support is in place.

MOBILIZATION STATION PHASE

7-32. This phase begins when the AC/RC unit arrives at the MS or mobilization site. The unit UMT is validated to meet deployment unit validation criteria resulting in assuring religious support mission capability. Necessary chaplain and chaplain assistant individual and collective training are conducted. Timeliness is critical. Integration to the Soldier Readiness Program (SRP) for preparation of overseas movement (POM) activities for religious support input is conducted.

PORT OF EMBARKATION PHASE

7-33. This phase begins with the arrival at the APOE or SPOE. Religious support is critical to ensure the forces are spiritually prepared for deployment and combat. Worship, counseling, and other religious support tasks are initiated. UMTs departing are ensured spiritual support.

DEPLOYMENT PHASE

7-34. In addition to routine religious support operations, the installation's TDA UMTs have the responsibility to plan and to prepare to sustain religious support operations on the Installation throughout the mobilization, deployment, and redeployment and demobilization process.

7-35. The senior UMTs of the deploying units and the TDA UMTs coordinate plans for religious support during deployment. Planning considerations include the following:

- The care of families and authorized civilians.
- Procedures for consolidating chapel programs.
- The close-out of chapels (if necessary).
- The accounting, turning-in, or securing of unit and installation property.
- Survivor and casualty assistance.

7-36. FORSCOM Regulation 500-3-2 (FORMDEPS II), Deployment Guide, provides guidance and assigns responsibilities for support of the deployment of units.

7-37. UMTs of non-deployed units are included in installation plans to support deploying units. A garrison support unit (GSU) may be available to augment the installation's UMTs during deployment and mobilization.

Along with the installation's non-deployed UMTs, they assist with religious support to families until the deployed forces return.

7-38. A thorough understanding and appreciation for the Transfer of Authority of deploying units to the theater commander is critical to the installation chaplain's mission. At a time designated in the deployment order, the chain of command shifts to the gaining unit, usually the theater commander. When that TOA occurs, the relationship of the deploying UMTs to the installation's UMTs changes correspondingly. This new relationship remains in effect until the unit redeploys.

SUSTAINMENT PHASE

7-39. To sustain the force projection platform, the installation UMT planner must consider all sustainment requirements for the non-deployed units, units at all deployment stages, and family members of both military and authorized civilians. The continued training and readiness posture is maintained for sustaining the forward religious support mission.

Sustainment: Family Readiness Groups

7-40. Command sponsored Family Readiness Groups (FRGs) are a vital link between soldiers, their families, the command, and the unit. Volunteer family members lead FRGs. Each deploying unit establishes its own FRG. Establishing the FRG prior to the deployment alert is extremely helpful in developing support plans and operational procedures for providing assistance to the unit' families. Chaplains and chaplain assistants support these groups.

Sustainment: Casualty Notification

7-41. Notification of the next of kin (NOK) is the commander's responsibility. Chaplains are not notification officers, but they may be a part of the team that makes the notification. In the event of a disaster or mass casualty situation, they may serve on the Casualty Assistance Team. Religious support to grieving families is a priority for UMTs and appropriate training in such religious support is a necessity.

REDEPLOYMENT AND DEMOBILIZATION PHASE

7-42. Religious support during the redeployment and demobilization phases consists of the spiritual and emotional preparation of units from an area of operation to follow-on designated CONUS or OCONUS bases. Redeployment religious support must be planned and executed to insure the proper allocation of resources of an installation. Planning must encompass all phases of an AC force returning to its home station, mobilized RC forces, and individuals through DMS back to their home stations.

7-43. This phase begins with the arrival of forces at the mobilization station/installation. The installation UMT supports the commander in receiving the forces, receiving sustainment equipment and supplies, and helps develop a reception and reconstitution or onward movement plan.

7-44. Demobilization is a critical factor for strategic reconstitution. The installation staff UMT plans and executes the commander's demobilization plan. Reunion support to soldiers, authorized civilians, and families are given to ease the transition from mobilization/deployment to demobilization/redeployment. Counseling, both individual and group, workshops, and worship are critical for reconstitution. Demobilization of some units may occur at the same time other units are being mobilized, deployed or redeployed. Religious support planning must encompass a possible ongoing planning and execution process.

7-45. Religious support plans encompass—

- Reunion briefing for family members and friends.
- Reunion support for returning units.
- Back briefing from returning and TDA UMTs.
- Religious support to garrison command and staff.
- Critical event debriefing.

DOMESTIC SUPPORT

7-46. 7-46. A presidential declaration of a major disaster or emergency usually precedes Domestic Support Operations (DSO). The installation staff UMT plans and executes the commander's domestic support mission. As in other operations, the UMT mission does not change. The installation becomes the force projection for Army forces to the area of operation. DOD Directive 3025.15, Military Assistance to Civil Authorities (MACA) provides directives and guidance in responding to natural and manmade disasters and includes military assistance to civil disturbance, counter drug activities, combating terrorism, and law enforcement. Installation UMTs must know their role and limitations in this type of operation. Installation UMTs support local, state, and federal civil authorities (see JP 3-07.7; JP 1-05; FM 100-20).

ROLE OF THE INSTALLATION UNIT MINISTRY TEAM

7-47. The Installation Chaplain and staff provide seamless religious support to the non-linear battlespace extending from the CONUS deployment and sustaining base to the forward edges of the battlespace. The religious support mission extends throughout the full spectrum of military operations. The community religious support mission encompasses all military units, activities, organizations, and authorized personnel.

7-48. The importance and influence of the Chaplaincy to the spiritual and moral health of the unit (to include the sustaining base or garrison) and in spiritual matters has been valued throughout the history of the Army. Commanders recognize the value of the chaplain and the total UMT both in combat and garrison.

RESPONSIBILITIES AND ACCOUNTABILITY

Staff Responsibilities

7-49. The Installation Chaplain provides support and advice to the commander and staff and to all installation agencies on all matters of religion, ethics/morals, and morale as affected by religion. The Installation Commander is responsible for the religious support and spiritual fitness of soldiers, family members, and authorized civilians. This includes insuring comprehensive worship opportunities that are part of the CMRP.

7-50. Army chaplains and chaplain assistants are the agents of trust representing the Army's values and the needs of the military in the commander's religious support program. Chaplains and chaplain assistants are required to provide religious support in all military units, activities, organizations, and communities.

CHAPLAIN (O-6)	*CHAPLAIN ASSISTANT NCO (E-7/ 8)*
Role Commissioned Officer/Staff Religious Professional Title 10 Protected	***Role*** Noncommissioned Officer Specialized Assistant Religious Support Specialist
Mission Advises commander on religious support and quality of life issues to support the unit's mission. Performs/provides/coordinates the religious support mission for the command. Installation Director of Religious Support	***Mission*** Advises Chaplain and NCO Chain. Provides specialized assistance in areas of religious support to support the chaplain. Perform Garrison Staff functions. Installation Religious Support NCOIC.
Parallel Tasks 1. Supervise, mentor and assess subordinate Chaplains. 2. Recommend personnel assignments. 3. Supervises the contingency, mobilization plans. 4. Responsible for religious support training and professional development.	***Parallel Tasks*** 1. Supervise, mentor and assess subordinate chaplain assistants. 2. Recommend Chaplain Assistant assignments. 3. Prepares the contingency, mobilization plans. 4. Plans and conducts religious support training.

7-51. Installation Chaplains have staff responsibility for the following:

- Development and execution of the Command Master Religious Program under the direction of the commander; which includes all worship services, memorial services, religious activities, professional training, religious education, spiritual fitness and moral leadership training, mobilization plans and contingency operation planning, and pastoral support for soldiers, families, and authorized civilians.
- Conduct of officer personnel management and coordination enlisted personnel actions for Unit Ministry Team (UMT) assignments and professional development.
- Development of an annual training plan for chaplain and chaplain assistant training.
- Development and execution of the installation commander's suicide prevention program.
- Chaplain Family Life Centers which provide comprehensive programs for training UMT members, and family programs, such as, marriage and family counseling and family enrichment programs.

- Administration and management of the installation non-appropriated chaplains' fund.
- Supervision of the use of chapels and religious facilities to include planning for new and renovated religious use facilities.
- The Installation Chaplain is the personal staff officer to the commander for the religious support portions of planning, programming, force structure actions, budget input, crisis management, and force protection.
- Provide the commander unit spiritual fitness assessments.
- Support the commander's METL through implementing a religious support installation METL.

7-52. The TDA force structure is required in the active and reserve component to maintain adequate military rotation and surge capability for mobilization and deployments and for military career progression of chaplains and chaplain assistants. These TDA positions are required to train and develop military religious support leaders in religious support skills for the full spectrum of military operations.

Relationship To The Chief Of Chaplains

7-53. The Installation Chaplain section implements the Chief of Chaplains (CCH) Strategic Plan and guidance on providing religious support across the full spectrum of operations from the power projection base to the theater of operations and return. They plan and execute the Chaplaincy mobilization and contingency plans. The installation staff UMT implements the CCH training strategy for individual UMT members and collectively. They train and prepare UMTs in support of the CCH guidance for the Army.

Relationship To Major Army Commands

7-54. Installations are part of Major Army Commands (MACOMs) or MSCs. The installation will coordinate, review, and recommend assignment of chaplains and chaplain assistants. They will support the MACOM or MSC staff chaplain contingency plans. Also, they will implement the religious program as set forth through the command. Again, as Transformation Installation Management takes effect, the role between Installation Chaplains and MACOMs will likely change. UMTs should keep abreast of such changes.

SUPERVISORY RESPONSIBILITIES

7-55. The assignment instructions received by every chaplain at the time of his/her receipt of permanent change of station (PCS) orders, contains the following statement: "Chaplain will perform and/or provide direct and general religious support in accordance with Unit Religious Support Plan and Command Master Religious Program." The installation staff UMT establishes staff supervision and implements the CMRP. They supervise the training of UMTs for the CCH in cooperation and coordination with tenant command UMTs. They supervise garrison religious support force structure for the commander, both active and civilian. On an installation or in a community, the staff chaplain has overall staff responsibility for the religious support mission. In fulfilling that responsibility, the staff chaplain will coordinate and use total UMT assets from both the TDA and TOE units. The installation chaplain assistant NCOIC advises the chaplain on chaplain assistant assignments, training, and other critical support issues.

SPECIALIZED RELIGIOUS SUPPORT

7-56. The Army is committed to supporting its families during routine operations and during deployment. Commanders employ the full range of installation resources to support families.

7-57. The provision of religious support to families is the responsibility of the installation chaplain. The Family Life UMT and the hospital UMT assist in their related areas in providing support. When not deployed, each unit's UMTs also have a responsibility to support their unit's families.

7-58. When units are deployed the RC will provide UMTs to assists in sustaining support to families. UMTs support the commander's Family Readiness Group (FRG) initiatives and also provide a vital role in supporting families when the unit's soldiers are deployed. Though not specifically a UMT responsibility, FRGs have a traditional relationship with the UMTs.

FAMILY LIFE UMT

7-59. The Family Life Chaplain's primary mission is to train UMTs and to direct the Chaplain Family Life Center (CFLC). Family life ministry consists of three major elements: enrichment, prevention, and intervention. These elements are expressed through educational programs, family wellness development programs, and marriage and family counseling.

7-60. The Family Life Chaplain consults with senior chaplains to determine the specific training needs for subordinate units. Each Family Life Chaplain tailors the training program to meet the needs of the installation and the mission of the tactical units. Training could be in the form of technical assistance, educational resources, and specific training for UMTs. In addition, the Family Life Chaplain helps UMTs prepare families for the stress of mobilization and deployment. This support is directly related to the installation's mission as the force projection platform. Each Family Life UMT supports the UMTs of the tactical units on the installation. They will tailor the training program to meet the needs of the installation and the mission of the tactical units on the installation. For example, a Family Life UMT will consult with a Division UMT to determine the division's specific training needs in family life issues. By training the Division's UMTs in family issues, the Family Life Chaplain serves as a direct link between the installation chaplain and the UMTs in sustaining the force. Training topics include—

- Ministry to FRGs.
- Family stress.
- Communication skills.
- Family counseling skills.
- Reunion after deployment.
- Grief and loss.

7-61. A healthy home environment directly influences the soldier's performance of duty. With its support of the family, the family life ministry program sustains the soldier's spiritual fitness in peacetime and during combat.

HOSPITAL UMT

7-62. The Army Medical Department preserves the fighting strength of the Army, maintains the health of soldiers, and supports family members with medical services.

7-63. Chaplains and chaplain assistants assigned to hospitals help patients and families deal with medical problems and crises. Hospital chaplains at all levels have special training and skills that equip them to do the following:

- Assist people with feelings of fear, loneliness, anger, failure, and loss of faith.
- Provide pastoral care for those suffering from critical or chronic illnesses or injuries.
- Prepare religious support contingency plans for mass casualty situations.
- Facilitate support groups for patients, their families, and staff.
- Serve as members of interdisciplinary case management teams and hospital committees.
- Provide pastoral care to hospital staff.
- Provide training for staff members and UMTs.

CONFINEMENT UMT

7-64. Most Army installations do not have local installation confinement facilities. Soldiers are usually held in confinement by special arrangement with local authorities. The installation staff chaplain provides pastoral care for locally confined soldiers in keeping with the guidance of the installation commander. The United States Disciplinary Barracks (USDB) is provided a UMT for RS. UMTs provide religious support for both prisoners and facility staff members. This religious support includes performing or providing for the following:

- Worship
- Religious education
- Counseling
- Moral leadership
- Spiritual fitness training
- Family support

GARRISON SUPPORT UNITS

7-65. Garrison Support Units (GSU's) come from the RC to support base operations (BASOPS) during continuing operations, and assist in other installation missions during Annual Training (AT) periods. Each GSU has UMT assets within their TDA. (See The Chief of Chaplains Crisis Management UMT Assets Handbook, October 2000.)

7-66. When installations are called upon to act as projection platforms, they require augmentation to meet their force projection responsibilities. The installation's increased religious support mission also requires additional support and augmentation. The scope of the installation's religious support mission encompasses support to soldiers remaining on the installation, soldiers processing through the installation; families of soldiers assigned to the installation, families of soldiers processing through the installation, and increased religious support to the installation hospital. The requirement for religious support augmentation will increase as the level of mobilization increases.

Training.

7-67. The supported Installation Chaplain will provide training to the GSU UMT personnel.

Organization

7-68. UMTs will be embedded in the GSU. The number of UMTs required varies from installation to installation. The number of UMTs is determined by applying an allocation rule to the number of projected personnel associated with the each installation's mobilization mission.

SUMMARY

7-69. The installation staff UMT is a critical part of the overall contingency plans of the commander. They provide a trained ready force of UMTs to meet the full spectrum operational mission of the U.S. Army. To adequately support the future requirements of an Army capable to conducting prompt and sustained operations, installation UMTs must plan and execute the religious support mission in support of the National Military Strategy, the OCCH (Office of the Chief of Chaplains) strategic plan, and the installation commander's CMRP. The installation staff UMT must—

- Support the UMTs assigned to war-fighting units.
- Focus on Army and Chaplaincy core competencies.
- Implement the command CMRP for readiness, mobilization, and deployment.
- Provide quality and effective training for UMTs.
- Maintain stewardship of religious support assets.
- Provide the projection base for force projection and quality ministry for families and tenant units.

RESOURCES

7-70. The following is a list of main resources utilized for writing and updating this doctrine.

- AR 165-1.
- JP 1-05.
- JP 5-00.2.
- JP 3-0 (100-5).
- JP 3-10.
- FM 16-1.
- FM 25-100.
- FM 25-101.
- FM 100-11.
- FM 100-9.
- FM 22-100.
- FM 100-20.

- FM 100-19.
- DA Pam 165-3.
- National Military Strategy
- Joint Vision 2010
- Army Vision 2010
- U.S. Army Chaplaincy Strategic Plan FY 2000-FY 2005
- Lessons Learned from past and current operation
- Lessons Learned from the training centers
- The Chief of Chaplain's UMT Crisis Assets Handbook, October 2000

Appendix A

RELIGIOUS SUPPORT IN CIVIL MILITARY OPERATIONS

Throughout U.S. history, chaplains have supported commanders through advisement on matters pertaining to the interface between Religious Support Operations and Civil Military Operations (CMO). This appendix provides the necessary guidelines for chaplains to engage in this role as required by their commanders. (See Appendix G for discussion on CD-C) UMTs are reminded that CMO support is a secondary responsibility and that the personal delivery of religious support is always the UMT's imperative.

ADVISING THE COMMANDER

A-1. Chaplains will support the commander through advisement in the following areas that may influence CMO:

- Indigenous religious, ethnic, and cultural influences.
- Beliefs, practices, and customs of religious groups in the AO.
- Religious issues related to displaced civilians.
- Human welfare needs as requested by religious NGOs and IOs.
- Relations with indigenous religious leaders when directed by the commander.

A-2. Under Title X of the U.S. Code, Chaplains should not perform the following:

- Direct participation in negotiations or mediations as sole participant.
- Human intelligence (HUMINT) collection and/or target acquisition.

PROVIDE AND PERFORM RELIGIOUS SUPPORT

A-3. Chaplains will provide worship and pastoral care in Civil Military Operations in accordance within the following guidelines:

- The priority of responsibility for Army Chaplains is to provide worship opportunities and pastoral care to soldiers and authorized civilians in the AO.
- Chaplains will provide for pastoral counseling, visitation, religious education, spiritual fitness training, moral leadership training, critical event debriefing and the provision of religious literature and items.
- Chaplains will identify and notify the command of meaningful opportunities for soldiers to participate in Civil Military Operations to promote the morale and spiritual well being of deployed soldiers.

- In coordination with the Civil Military Operations Center, chaplains may provide certain religious support to refugees and displaced persons when directed by the Commander, and after consultation with Staff Judge Advocate (SJA). In such cases, it is critical to avoid any activities that can be construed as proselytizing among refugees or displaced persons.
- Every effort to include indigenous religious leadership to meet these needs should be taken.
- *CMO and RS tasks will need to be coordinated to ensure success.*

CMO Tasks	Religious Support Tasks
Advise commander of legal obligations and moral considerations.	Advise commander of moral considerations as affected by religion.
	Advise command on accommodating religious practices in AO.
Establish policy for the handling of dislocated civilians.	Monitor care of displaced civilians to ensure it is moral/ethical.
Plan displaced civilian operations.	
Coordinate displaced civilian operations.	Advise command on religious support plan for displaced civilians.
Provide displaced civilian support.	
Plan humanitarian/civic assistance activities.	Plan religious support activities. Assist CMO personnel in coordinating with local religious leaders for implementation of CMO activities.
Provide public welfare support.	Assist command in meeting human welfare needs in conjunction with religious NGOs & IOs –clothing, food shelter.
Coordinate health, welfare, and morale services.	
Provide civilian supply support.	
Provide food and agriculture support.	
Provide information on cultural considerations.	Advise leadership on beliefs, practices, and customs of religious groups in AO
Provide cultural affairs support.	Advise CMO personnel on the role/influence of religion in cultural affairs.
	Assess distinct religious, cultural, ethnic, and gender ministry needs
	Brief the commander of religious influences on contingency operations
Provide language support.	
Provide arts, monuments and archives support.	Advise command on impact of local religions, sites and facilities which affect the mission
	Advise command on the training of soldiers with respect to religious beliefs, practices, sites and facilities

CMO Tasks	Religious Support Tasks
Provide legal support.	
	Implement religious support in a Combined and/or Joint Area of Operation with Consid-eration to Host Nation Religious Background.
Prepare an area assessment.	Performs analysis of local religion as it affects mission.
Prepare an area study.	Prepares the RS annex to the OPORD, which includes impact of indigenous religions on AO and mission.
Implement the command map program.	Provide input to CMO personnel regarding local religions.
CMO must analyze the effect of FNS on the morale of U.S. soldiers and on the psychological condition of the local populace.	Advise leadership on negative changes in solder attitude on local population.
Plan disaster relief operations.	Provide religious support for humanitarian assistance survey team (HAST) operations.
Plan noncombatant evacuation operations.	
Commanders must segregate civilians from EPWs and civilian internees to protect them as required by interna-tional law.	Coordinate religious support for Enemy Pris-oners of War.
Coordinate operations.	Coordinate religious support activities with CMO personnel and local religious leaders.
Perform command functions.	Provide religious support to CMO personnel.

Appendix B

COMBAT AND OPERATIONAL STRESS CONTROL AND RELIGIOUS SUPPORT

References: FM 8-51.
FM 22-51.
Textbook of Military Medicine: War Psychiatry, Part I; Warfare, Weaponry, and the Casualty Office of the Surgeon General of the United States Army, Falls Church, Virginia: 1995

INTRODUCTION

This appendix addresses the general role of the Unit Ministry Team in the commander's program for Combat and Operational Stress Control. Unit Ministry Teams, imbedded within units down to battalion level, provide immediate support to leaders in fulfilling their battle fatigue responsibilities. Unit Ministry Teams also assist in training leaders to recognize battle fatigue identification and intervention responsibilities. In cooperation with unit medical personnel, Unit Ministry Teams serve as a primary referral agency to mental health resources.

BACKGROUND

B-1. The negative impact of Combat and Operational Stress has been known by many names throughout the history of warfare: Nostalgia, Soldier's Heart, Shell Shock, War Neurosis, Combat Stress Reaction, and Battle Fatigue. Combat Stress Behaviors cover the full range of behavior in combat from the highly positive to the completely negative.

B-2. Positive Combat Stress Behaviors include heightened alertness, strength, endurance, and tolerance of discomfort. Examples of positive Combat Stress Behaviors would include the strong personal trust, loyalty, and cohesiveness, which develops among, peers in small units, as well as between leaders and subordinates.

B-3. Dysfunctional Combat Stress Behaviors come in two forms: Misconduct Stress Behaviors and Criminal Acts, and Battle Fatigue.

B-4. Misconduct Stress Behaviors and Criminal Acts range from simple UCMJ infractions to outright criminal acts such as killing enemy prisoners of war, threatening or killing superiors, or looting, pillaging, and rape.

B-5. Battle Fatigue, also referred to as Combat Stress Reaction or Combat Fatigue, is the distress and impaired performance resulting from the accumulated stressors of the combat or operational situation. Signs of Battle Fatigue would include hyper alertness; fear and anxiety; grief, self-doubt, and guilt; loss of confidence and hope; impaired duty performance; erratic actions and outbursts; freezing and immobility; terror, panic, and flight; and impaired speech, vision, touch, and hearing. In extreme cases Battle Fatigue may result in paralysis and hallucinations.

THE BASICS OF TREATMENT

B-6. The principles for stabilizing treatment of soldiers experiencing Battle Fatigue are summarized by the acronym **PIES**.

- **Proximity**: Battle Fatigue cases are treated as far forward as tactically possible and in close proximity with the soldier's unit with the goal of returning the soldier to duty after a period of several days.
- **Immediacy**: Immediate initiation of treatment.
- **Expectancy**: Expectation of rapid and full recovery and return to duty.
- **Simplicity**: Simplicity of approach, according to the four Rs:
  - *Reassurance*: Reassurance of normality.
  - *Rest*
  - *Replenishment*: Replenishment of nutrition, hydration, hygiene and sense of physical well being.
  - *Restoration*: Restoration of confidence through talk and activities.

B-7. It is important to note that the soldier's identity as a soldier is maintained and reinforced. Throughout treatment Battle Dress Uniforms are worn, not hospital pajamas. Rank distinctions and military courtesy are maintained, and these soldiers are responsible for self-care and helping others.

RELIGIOUS SUPPORT AND COMBAT/OPERATIONAL STRESS CONTROL

B-8. Soldiers' inner resources are generally rooted in their religious and spiritual values. In combat, soldiers often show more interest in their religious beliefs. When religious and spiritual values are challenged by the chaos of combat, soldiers may lose connection with the inner resources that have sustained them. Then they become targets of fear, and hopelessness, and eventually fall victim to Battle Fatigue. Further, they become at risk for Misconduct Stress Behaviors. The Unit Ministry Team is the primary resource available to soldiers experiencing these dilemmas and seeking to refocus their spiritual values.

The Unit Ministry Team's Role

B-9. Unit Ministry Teams provide preventative, immediate, and restorative spiritual and emotional support and care to soldiers experiencing Battle Fatigue.

B-10. Preventative Religious Support. The Unit Ministry Team assists in preventing Battle Fatigue and Misconduct Stress Disorders through Spiritual Fitness training. Ministry of Presence with soldiers, assigned DA civilian and contractor personnel is critical. The Unit Ministry Team provides a stabilizing influence on personnel, and assists them in strengthening and regaining personally held spiritual values. Preventative activities include—

- Worship opportunities.
- Private and group prayer opportunities.

- Religious literature and materials.
- Scripture readings.
- Sacraments and ordinances.
- Opportunities for working through frustration, fear, anxiety, and anger.
- Assistance to personnel and families prior to deployment, emphasizing family strengths.

B-11. Immediate Religious Support. The Unit Ministry Team assists commanders in the identification of personnel experiencing negative reactions to combat and operation stress: Battle Fatigue and Misconduct Stress Behaviors. The Unit Ministry Team works closely with the unit's leaders and medical personnel to care for Battle Fatigue cases through religious support and comfort. Immediate Religious Support activities may include—

- Presence with the soldier.
- Conversation focused upon fears, hopes, and other feelings.
- Conversation focused upon forgiveness.
- Prayer with personnel.
- Prayer for fallen comrades and Memorial Ceremonies and Services.
- Rites, sacraments, and ordinances, as appropriate.
- Sacred Scripture.

B-12. Restorative Religious Support. Following an operation, a unit may require reconstitution. Surviving soldiers may need to rebuild emotional, psychological, and spiritual strength. Depending upon the spiritual, emotional, and physical condition of the unit's soldiers, the organic Unit Ministry Team may need augmentation from higher echelons or other units. Restorative Religious Support activities may include—

- Worship, sacraments, rites, and ordinances.
- Memorial ceremonies and services.
- Religious literature and materials.
- Grief facilitation and counseling.
- Reinforcement of the soldiers' faith and hope.
- Opportunities for soldiers to talk about combat experiences and to integrate those experiences into their lives.

COMBAT STRESS BEHAVIORS

- POSITIVE COMBAT STRESS BEHAVIORS
- DYSFUNCTIONAL COMBAT STRESS BEHAVIORS
  - MISCONDUCT STRESS BEHAVIORS AND CRIMINAL ACTS
  - BATTLE FATIGUE

POSITIVE COMBAT STRESS BEHAVIORS	MISCONDUCT STRESS BEHAVIORS AND CRIMINAL ACTS	BATTLE FATIGUE
UNIT COHESION	MUTILATING ENEMY	HYPERALERTNESS
LOYALTY TO BUDDIES	DEAD	FEAR, ANXIETY
LOYALTY TO LEADERS	NOT TAKING PRISONERS	IRRITABILITY, ANGER, RAGE
IDENTIFICATION WITH	KILLING ENEMY	GRIEF, SELF-DOUBT, GUILT
UNIT TRADITION	PRISONERS	PHYSICAL STRESS
SENSE OF ELITENESS	KILLING	COMPLAINTS
SENSE OF MISSION	NONCOMBATANTS	INATTENTION,
ALERTNESS AND	TORTURE, BRUTALITY	CARELESSNESS
VIGILANCE	KILLING ANIMALS	LOSS OF CONFIDENCE
EXCEPTIONAL STRENGTH	FIGHTING WITH ALLIES	LOSS OF HOPE AND FAITH
AND ENDURANCE	ALCOHOL AND DRUG	DEPRESSION, INSOMNIA
INCREASED TOLERANCE TO	ABUSE	IMPAIRED DUTY
HARDSHIP	RECKLESSNESS,	PERFORMANCE
DISCOMFORT	INDISCIPLINE	ERRATIC ACTIONS,
PAIN	LOOTING, PILLAGE, RAPE	OUTBURSTS
INJURY	FRATERNIZATION	FREEZING, IMMOBILITY
SENSE OF PURPOSE	EXCESSIVELY ON SICK	TERROR, PANIC RUNNING
INCREASED FAITH	CALL	TOTAL EXHAUSTION
HEROIC ACTS	NEGIGENT DISEASE,	APATHY
COURAGE	INJURY	LOSS OF SPEECH SKILLS
SELF-SACRIFICE	SHIRKING, MALINGERING	AND HEARING
	COMBAT REFUSAL	WEAKNESS, PARALYSIS
	SELF0INFLICTED	HALLUCINATIONS
	WOUNDS	DELUSIONS
	THREATENING/KILLING	
	OWN LEADERS	
	("FRAGGING")	
	ABSENCE WITHOUT	
	LEAVE	
	DISERTION	

Appendix C

CRISIS INTERVENTION MANAGEMENT AND CRITICAL EVENT DEBRIEFINGS

BACKGROUND

There are currently three major schools of Crisis Intervention Management, of which Critical Event Debriefing (CED) by a variety of names, are a part. These are the International Critical Incident Stress Foundation's Critical Incident Stress Management (CISM), the National Organization for Victim Assistance (NOVA) Group Crisis Intervention (GCI), and the American Red Cross Disaster Response. In each case, CED is a component of the larger Crisis Intervention model, and training and certification requirements vary from model to model.

PURPOSE

C-1. Crisis Intervention helps small units soon after exceptionally traumatic events to—

- Quickly restore unit cohesion and effectiveness.
- Reduce short-term emotional and physical distress.
- Mitigate long-term distress and "burnout".
- Facilitate the transition from traumatic event (s) to a sense of normalcy and aids in future transitions.

REQUIREMENT

C-2. Leaders of small units should ordinarily conduct an after action review (AAR) when a mission is completed. After an exceptionally distressing mission or event, when METT-TC dependent, unit leaders may coordinate crisis intervention. Members of the Unit Ministry Team are equipped to provide this intervention.

C-3. Situations that may warrant crisis intervention include—

- Death of the unit member.
- Death or suffering of noncombatants (especially women and children).
- Handling of human remains, management of carnage, extensive disaster devastation.
- Friendly fire incident.
- Situation involving serious error, injustice, or atrocity.

C-4. Crisis intervention may be conducted by the Unit Ministry Team or may be coordinated with Mental Health/Combat Stress Control or other medical personnel.

PERSPECTIVE

C-5. Participants in crisis intervention are normal people who have survived an abnormal situation. CED is a generic term for a specific tool within the broader practice of crisis intervention management. CED is neither therapy nor clinical counseling. Rather, it is basic and wise interventive maintenance for the human spirit.

THE CRITICAL EVENT DEBRIEFING

C-6. WHO

- A functional organization: crew, team, squad, or platoon.
- Strangers, civilians, family members thrown together into a traumatic situation by chance.
- Normally includes only those directly involved in the event.
- Chain of Command included only if involved in the event.
- Includes members of identified Crisis Intervention Team.
- Media and outsiders excluded.

C-7. WHAT

Normally, the team consists of a debriefing leader and an assistant, with one additional assistant for every 10 people in a group of more than 20.

C-8. WHEN

- Most effective when conducted 12-72 hours after the event.
- Allow for enough rest and recovery for all to be alert and involved.
- After completing an operation with no expected intense follow-on operations for next 72 hours.
- Expected duration: 2-3 hours.
  When ongoing tactical operations prohibit CED, a briefer "defusing debriefing" focusing on cognitive reactions to the situation can be substituted

C-9. WHERE

- An emotionally neutral place, relatively safe from enemy action, distraction, observation.
- Reserve position or assembly area.
- With enough light to see all participants.
- Shelter from bad weather.

C-10. HOW

- **Introductory Phase**
  Purpose: To introduce the Crisis Intervention Team and explain the process.
  Chaplain or Leader, explains the grounds rules:
  - Personal information or feelings shared in the group during CED are not to be shared outside the group

---

Note: This does not override legal or moral responsibility to report violations of UCMJ or the Law of Land Warfare.

  - No notes or recordings.

- No breaks are scheduled, but anyone may leave as needed, returning as soon as possible.
- No one is expected to speak after the Fact Phase.
- Each participant speaks for self, not others.
- Everyone is equal during CED. All ranks speak frankly, with proper courtesy.
- CED is not an AAR, but a discussion to clarify what happened and to restore sense of well-being.
- Fact-finding, not Fault finding.
- "Facts" include personal reactions the event.
- The Team is available following the CED.

- **Fact Phase**
  Purpose: Reconstruct the event in detail, in chronological order, as an unbroken "historical time line" viewed from all sides and perspectives.
  Chaplain/Leader encourages all participants to start their stories before the event occurred and to work up to it.
  Participants—
  - The first person involved in the event is asked to tell his version of the story- how it started, his role (duty position), and what he saw, heard, smelled, and did-step by step.
  - One by one, participants are drawn into the first person's story and then asked to describe their observations and actions.
  - The Chaplain/Leader encourages participation. Everyone is asked to speak.
  - When disagreements over what occurred arise, the Chaplain/leader elicits group observations to clarify memories.
- Thought Phase
  Purpose: To personalize the event.
  Chaplain/Leader facilitates transition from facts to the personal.
  Participants are asked to share—
  - "What were you thinking as the event started?"
  - "What went through your mind when you saw, smelled, or did...?"
  - "What was your first thought when you came off 'automatic' or 'autopilot'?"
- **Reaction Phase**
  Purpose: Identify and ventilates feelings (emotions) raised by the event.
  Chaplain/Leader emphasizes that all emotional reactions deserve to be expressed, respected, and listened to.
  Participants are encouraged to share "reactions":
  - "What was the worst thing about the event?"
  - "How did you react when it happened?"
  - "How are you feeling about that now?"
  - "If you could change one thing about the event, what would it be?"

  Chaplain/Leader listens for common themes, feelings, and misperceptions—
  - Feelings of anger at others for not helping.
  - Blaming self or others for things beyond one's control.

  - Feeling changed, different, worse than everyone else, cut off from others.

  Participants are asked to describe physical reactions to the event. Common symptoms include—

  - Gastro-intestinal distress
  - Frequent urination
  - Loss of bowel and bladder control
  - Loss of sexual interest
  - Heart pounding
  - Shortness of breath
  - Muscle, back, neck, and head ache
  - Trembling, jumpiness, and startle reactions

- **Reframe Phase**

  Purpose: To transition from emotional reaction to the cognitive.
  Chaplain/Leader facilitates the search for meaning in the event.
  Participants are encouraged to derive personal meaning in the event.

  - "What lessons could be learned from this event?"
  - "What is something positive that you will take away from this experience?"
  - "What good can be found in this tragic situation?
  - "What are you proud of in this event?"

- **Teaching Phase**

  Purpose: To reassure by teaching the participants that feelings and stress symptoms are normal reactions to abnormal conditions. Symptoms may last a while, but can be expected to resolve in time. If they don't resolve in time, seeking professional help is advised.
  Chaplain/Leader summarizes the thoughts, feeling, and symptoms expressed by the group, reemphasize normality; and reduce feelings of uniqueness, weakness, or injury.
  No predictions or glamorization of long-term disability should be given.
  The Unit Ministry Team may schedule additional training in stress management, coping strategies, grief process, and anger management at later dates.

- **Reentry Phase**

  Purpose: Complete and close the CED.
  Chaplain/Leader gives final invitation for comments and makes a summary statement. Follow-up resources are discussed.
  It is important for Unit Ministry Team personnel to make themselves available for follow-on conversations with individual participants following the CED.

Appendix D

INSTRUCTIONS FOR USING THE MODEL RELIGIOUS SUPPORT MATRIX

GENERAL INFORMATION (ADAPTED FROM FM 101-5)

This appendix provides a sample Religious Support Matrix. The matrix provides a highly visible, clear method to ensure all religious support providers are synchronized. It shows the relationships between activities, units, support functions and key events. Although the unit order will not appear in matrix format, the staff can write their annex as an execution matrix. The religious support appendix as part of the Service Support annex can be done in matrix format. An execution matrix depicts when and where specific supporting actions must occur.

HEADER INFORMATION

D-1. *Classification.* Army Regulation 380-5 contains the detailed description of classification procedures. If the entire plan is unclassified, no classification is required on the appendix.

D-2. *Annexes, Appendixes, Tabs, and Enclosures.* FM 101-5, 31 May 1997, p. H-7ff gives details on how annexes, appendixes, tabs, and enclosures are used. Follow unit SOP and include a matrix as shown in the sample in Figure D-1. The matrix should be included as an appendix to the Service Support Annex. An Arabic number designates the appendix; a capital letter designates the Annex . The Service Support Annex should include the Religious Support Matrix in the list of appendixes along with the other appendixes, such as Personnel and Legal.

D-3. *References.* References list information related to the matrix. The information should help the intended users understand the matrix. Map sheet series numbers may be included. The unit SOP does not need to be referenced. (Cf. FM 101-5, 31 May 1997 p. H-14)

D-4. *Time Zone.* The time zone used throughout the order is placed here. Operations across several zones use ZULU time. If the local time is chosen, the appropriate time zone suffix is used. The time zone is not based on geographical location but rather on the relationship to ZULU time. For example, Central Standard Time would normally be SIERRA. However, during daylight savings time ROMEO would be the appropriate suffix.

D-5. *DTG.* Date time group (DTG) refers to the time of signature in day, hour with suffix, month, and year format. E.g. 260900S JAN04 would be 26 January 2004 at 0900 hr in the SIERRA time zone.

D-6. *Task Organization*. Task Organization is found in Annex A of the OPORD/OPLAN. The UMT coverage plan is made by matching the UMTs with the units listed in the Task Organization. The unit designator like "Iron Horse" goes under the Unit name e.g. (2BCT). List the chaplain and assistant who provide the coverage. When the units are Task Force organized the chaplain of the Task Force is automatically responsible for all the units that would show up in their level Task Organization. If the coordinated Religious Support Plan calls for chaplains to cover units that are not a part of their unit's Task Organization for the entire operation, note those exceptions under the units listed to the right. If the exceptions change by phase, list the coverage in the concept of operation under units. List the chaplain and assistant who provide coverage and indicate the denomination or faith group of chaplains to more completely track coverage needs in the unit AO.

- Enemy Situation. Normally a reference to the Intelligence Annex will suffice. This presupposes that the chaplain has coordinated with the S2/G2 during the Religious Area Analysis (RAA) phase of planning. If there is anything specific about the enemy related to the RS mission that is not listed in the Intelligence annex, then it would go here.
- Friendly Situation. Normally, the basic order can be referenced here. Significant religious issues for the friendly forces are also added here if necessary. Note Religious Preference Profiles (RPPs) that differ from the norm. National Guard units can at times take local flavors. For instance, a National Guard unit in Utah may have a high percentage of soldiers who are members of the Church of Jesus Christ of Latter Day Saints. The Hawaii National Guard might have a unique local flavor as well. Anything that affects your religious support mission that is not the norm is noted here.
- Mission. State the Religious Support tasks and their purpose. Mission is to ___________ in order to _______________.
- Concept of Operations.
  - Location. Location answers the question, "Where will people go for services." During the different phases as the senior chaplain changes location people may have to go different places in order to receive or coordinate for Religious Support Services. Grid designators, Area of Operation, ISB chapel, are types of information that you can use.
  - Units. Note here exceptions to the coverage plan from what is listed under Task Organization. The purpose of the block is to state what UMT is providing which unit service. If the aviation chaplain cover the smoke platoon in the first phase but the field artillery chaplain covers the smoke platoon in the second phase, then list it here.
  - Time Frame. The time that is referred to here is the time schedule for services rendered. This is not the beginning and the ending of the phase. The start and ending of a phase can be listed under the Title of the Phase. Are there any special services or special religious holidays that happen during the operation? These can be listed here. At battalion level you may be able to state when you are going to hold a service for a particular company.

- UMT Priority of Support. UMT Priority of support refers to the services rendered by the UMT. While the unit is in the ISB the priority might be generally nurture the living. Then you would want to spell out just what that means: religious services, counseling opportunities, classes on stress, etc. During the battle the priority might be caring for casualties in general. Then what does that mean specifically? You could list praying for casualties, ministering to care givers, etc. During the post-battle phase honoring the dead might be the priority. During redeployment nurturing the living might include reunion briefings.
- Priority of Support. Priority of Support refers to the units that will receive the services offered by the UMT. While in the ISB, you may give priority to the scouts because of the danger of their mission and because they will be hard to reach later. Often, the religious support priority matches the commander's priority of support as found in the order. Normally, the main effort receives the priority of support.

D-7. *Coordinating Instructions*. Coordinating Instructions tells the unit how they will coordinate for services. How do people coordinate for religious services? How does the unit coordinate for Roman Catholic coverage or some other minority faith need? Before chaplains go on a civil affairs type of mission what coordination needs to take place. How are extraordinary Ministers of the Eucharist being used? How will denominational service leaders (DSLs) function or will they?

D-8. *Service Support*. State how many days of supply each UMT will need. State how re-supply is conducted. Normally, re-supply is through S4 channels.

D-9. *Command*. Report times and formats can be listed here. You can list the specific chain of succession for this operation. You should have the general procedures for the chain of succession in your TSOP.

D-10. *Signal*. List primary and alternate means of communication. List specific information that is needed to contact the chaplain who is responsible for the appendix. If it is a Brigade appendix it is the brigade chaplain. If it is a battalion appendix, then it should be the battalion chaplain.

D-11. *Safety*. List specific safety concerns. Could state that the UMTs need to do a Risk Assessment before certain types of missions (e.g. Civil Affairs or movement to new locations).

D-12. Extraneous Information.

- Name and position example Dolinger, TF CH
- Time Zone: The time zone is the same one used throughout the order.
- OFFICIAL: If the commander signs the original this is not used. The signed copy is a historical record and should remain in headquarters files. The preparing officer signs under OFFICIAL when the commander does not sign it.
- Types of services available
- Designation and location of unit or place providing the service.
- Schedules for services

- Specific missions for units (Special religious groups or holidays)
- Concept of operations.
  - Priority of effort
  - Location
  - Procedures
  - Policies
  - Requirements
  - Phases are specific parts of an operation that differ from those that precede or follow. Phases are indicated by time, distance, terrain, or event occurrences. These will be identified in the order.

CLASSIFICATION

APPENDIX 4 (RELIGIOUS SUPPORT MATRIX) TO ANNEX I (SERVICE SUPPORT) to OPLAN 10-01 OPERATION OVERLORD

REF: AR 165-1, JP 1-05, FM 1-05, RB 16-100					TIME ZONE: ZULU		DTG: 060300JUN44	
TASK ORGANIZATION: See Annex A (Task Org) for each phase	**Unit Designator** CH Goode Assistants SSG Nock	**Subordinate Unit Designator** CH Poling Assistants SSG Bizarri	**Subordinate Unit Designator** CH Fox Assistants SGT Woods	**Subordinate Unit Designator** CH Washington Assistants SPC Duff	**Subordinate Unit Designator** CH Watters Assistants SGT Titus	**Subordinate Unit Designator** CH Rayburn Assistants SGT Calvin	**Subordinate Unit Designator** CH Spurgeon Assistants SGT Luther	**Subordinate Unit Designator** CH Augstine Assistants PVT Delta

1. SITUATION — **1a. ENEMY SITUATION:** Annex B (Intelligence) — **1b. FRIENDLY SITUATION:** Basic Order

2. MISSION: Provide RS to dispersed elements of BCT NLT 081100JUN44 and provide RS to MASCAL sites as directed.

3. EXECUTION: *(intent)*

3a. Concept of Operations	Phase I Title of Phase	Phase II Title of Phase	Phase III Title of Phases	Phase IV Title of Phase
Location	UMTs cover units starting w/ those closest to the front	UMTs locate at Casualty Collection Points	UMTs cover CCPs and units first to attack	TBD
Units	EN UMT covers smoke platoon	FA UMT covers smoke platoon	TBD	TBD
Time Frame	Average 5 services daily per UMT	Average 1 service daily plus personal casualty ministry	Average 2 services daily	TBD
UMT Priority of Effort	Religious services, Prayer, Counseling	Prayer for casualties and dying, ministry to care givers	Prayer for casualties, religious services	TBD
Priority of Support	IN TF, EN BN	IN TF, AR TF	EN BN, FA BN	TBD

3b. COORDINATING INSTRUCTIONS.

(1) All UMTs going on CMO missions must coordinate w/ the BDE UMT.

(2) Units desiring Roman Catholic Coverage must coordinate w/ the BDE UMT.

(3) UMTs will report service times IAW SOP.

4. SERVICE SUPPORT

(1) UMTs will carry 30 days of supply.

(2) Resupply is through unit S4 channels.

5. COMMAND AND SIGNAL

5a. COMMAND.

(1) Chain of succession is FSB, EN, FA UMTs

(2) BDE UMT will use TOC as primary location.

(3) Send in reports IAW TACSOP

5b. SIGNAL.

(1) Primary means of communication is FBCB2, DVNT, VIASAT, BDE Admin/log FM net, Messanger

(2) Radio suffix for all Chaplains is 9, expander for Chaplain Assistants is 0.

5c. SAFETY.

UMTs will conduct Risk Assessment prior to all convoy movement and CMO missions.

ACKNOWLEDGE:
OFFICIAL: MOSES, BDE Chaplain

NAME (HYATT)
Rank (COL)

Figure D-1. Sample Model Religious Support Matrix

Appendix E

UNITED STATES CODES RELEVANT TO RELIGIOUS SUPPORT OPERATIONS

This appendix contains extracts from the USC relevant to RS. The extracts range from general military law to laws governing conscientious objectors, accommodation of religious practices and Title 18, Posse Comitatus Act.

E-1. Section 3073. Chaplains. There are chaplains in the Army. The Chaplains include—

- The Chief of Chaplains.
- Commissioned officers of the Regular Army appointed as chaplains.
- Other officers of the Army appointed as chaplains in the Army.

E-2. Section 3547. Duties: chaplains; assistance required of commanding officers—

- Each chaplain shall, when practicable, hold appropriate religious services at least once on each Sunday for the command to which he is assigned, and shall perform appropriate religious burial services for members of the Army who die while in that command.
- Each commanding officer shall furnish facilities, including necessary transportation, to any chaplain assigned to his command, to assist the chaplain in performing his duties.

E-3. Section 3581. Command: chaplains. A chaplain has rank without command.

E-4. Section 774. Religious apparel: wearing while in uniform—

- General Rule. - Except as provided under subsection (b), a member of the armed forces may wear an item of religious apparel while wearing the uniform of the member's armed force.
- Exceptions. - The Secretary concerned may prohibit the wearing of an item of religious apparel.
  - In circumstances with respect to which the Secretary determines that the wearing of the item would interfere with the performance of the member's military duties; or
  - If the Secretary determines, under regulations under subsection (c), that the item of apparel is not neat and conservative.

- Regulations. The Secretary concerned shall prescribe regulations concerning the wearing of religious apparel by members of the armed forces under the Secretary's jurisdiction while the members are wearing the uniform. Such regulations shall be consistent with subsections (a) and (b).
- Religious Apparel Defined. In this section, the term "religious apparel" means apparel the wearing of which is part of the observance of the religious faith practiced by the member.

E-5. Section 6031. Chaplains: divine services—

- An officer in the Chaplain Corps may conduct public worship according to the manner and forms of the church of which he is a member.
- The commanders of vessels and naval activities to which chaplains are attached shall cause divine service to be performed on Sunday, whenever the weather and other circumstances allow it to be done; and it is earnestly recommended to all officers, seamen, and others in the naval service diligently to attend at every performance of the worship of Almighty God.
- All persons in the Navy and in the Marine Corps are enjoined to behave themselves in a reverent and becoming manner during divine service.

E-6. Section 8547. Duties: chaplains; assistance required of commanding officers—

- Each chaplain shall, when practicable, hold appropriate religious services at least once on each Sunday for the command to which he is assigned, and shall perform appropriate religious burial services for members of the Air Force who die while in that command.
- Each commanding officer shall furnish facilities, including necessary transportation, to any chaplain assigned to his command, to assist the chaplain in performing his duties.

DEPARTMENT OF DEFENSE DIRECTIVES

DODD 1300.6 – Conscientious Objectors

E-7. References:

- DOD Directive 1300.6, subject as above, May 10, 1968 (hereby canceled)
- DOD Directive 1332.14, “Administrative Discharges, “ December 20, 1965
- Section 6(j) of the Universal Military Training and Service Act, as amended (50 U.S.C. App. 456(j))
- Section 3103, title 38, United States Code

Definitions

E-8. **3.1. Conscientious Objective 1-0**. A firm, fixed and sincere objection to participation in war in any form or the bearing of arms, by reason of religious training

E-9. **3.1.1. Class1-O Conscientious Objector**. A member who, by reason of conscientious objection, sincerely objects to participation of any kind in war in any form.

E-10. **<u>3.1.2. 1-A-O Conscientious Objector</u>**. A member who, by reason of conscientious objection, sincerely objects to participation as a combatant in war in any form, but whose convictions are such as to permit a Military Service in a non-combatant status.

E-11. **3.3.** Non-combatant service or non-combatant duties (1-A-O) used interchangeably herein)

E-12. **3.3.1.** Service in any unit of the Armed Forces which is unarmed at all times.

E-13. **3.3.2.** Service in the medical department of any of type Armed Forces, wherever performed.

E-14. **3.3.3.** Any other assignment the primary function of which does not require the use of arms in com bat provided that such other assignment is acceptable to the individual concerned and does, not require him to bear arms or to be trained in their use.

E-15. **3.3.4.** Service aboard an armed ship or aircraft or in a combat zone shall not be considered to be combatant duty unless the individual concerned is personally and directly involved in the operation of weapons

E-16. **<u>3.4. Non-combatant Training</u>**. Any training which is not concerned with the study, use or handling of arms or weapons.

E-17. **5.3.1.** In order to find that an applicant's moral and ethical beliefs are against participation in war in any form and are held with the strength of traditional religious convictions, the applicant must show that these moral and ethical convictions, once acquired, have directed his life in the way traditional religious convictions of equal strength, depth and duration have directed the lives of those whose beliefs are clearly found in traditional religious convictions. In other words, the beliefs are clearly found in traditional religious convictions. In other words, the beliefs upon which conscientious objection is based must be the primary controlling force in the applicant's life.

E-18. **5.3.2.** A primary factor to be considered is the sincerity with which the belief is held. Great care must be exercised in seeking to determine whether asserted belief is honestly and genuinely held. Sincerity is determined by an impartial evaluation of the applicant's thinking and living in its totality, past and present. Care must be exercised in determining the integrity of belief and the consistency of application. Information presented by the claimant should be sufficient to convince that the claimant's personal history reveals views and actions strong enough to demonstrate that expediency or avoidance of military service is not the basis of his claim.

E-19. **5.3.2.1.** Therefore, in evaluating applications the conduct of applicants, in particular their outward manifestation of the beliefs asserted, will be carefully examined and given substantial weight.

E-20. **5.3.2.2.** Relevant factors that should be considered in determining an applicant's claim of conscientious objection in the home and church; general demeanor and pattern of conduct; participation in religious activities; whether ethical or moral convictions were gained through training, study, contemplation, or other activity comparable in rigor and dedication to the processes by which traditional religious convictions are formulated; credibility of the applicant; and credibility of persons supporting the claim.

E-21. **5.3.3.1.** Church membership or adherence to particular theological tenets are not required to warrant separation or assignment to non-combatant training and service for conscientious objectors.

E-22. **5.3.3.2.** Mere affiliation with church or other group which advocates conscientious objection as a tenet of its creed is not necessarily determinative of an applicant's position or belief.

E-23. **5.3.3.3.** Conversely, affiliation with a church or group which does not teach conscientious objection does not necessarily rule out adherence to conscientious objection beliefs in any given case.

E-24. **5.3.3.4.** Where an applicant is or has been a member of a church, religious organization, or religious sect, and where his claim of conscientious objection is related to such membership, inquiry may properly be made as to the fact of membership, and the teaching of the church, religious organization, or religious sect, as well as the applicant's religious activity. However, the fact that the applicant's may disagree with, or not subscribe to, some of the tenets of his church does not necessarily discredit his claim. The personal convictions of each individual will be controlling so long as they derive from moral, ethical or religious beliefs.

E-25. **5.4.** The burden of establishing a claim of conscientious objection as grounds for separation or assignment to non- combatant training and service is on the applicant. To this end, he must establish clear and convincing evidence (1) that the nature or basis of his claim comes within the definition of and criteria prescribed herein for conscientious objection, and (2) that his belief in connection therewith is honest, sincere and deeply held. The claimant has the burden of determining and setting forth the exact nature of his request, i.e., whether for separation based on conscientious objection (1-O), or, for assignment to non-combatant training and service based on conscientious objection (1-A-O).

E-26. **6.3.** The applicant shall be personally interviewed by a chaplain who shall submit a written opinion as to the nature and basis of the applicant's claim, and as to the applicant's sincerity and depth of conviction. The chaplain's report shall include the reasons for his conclusions. This opinion and report will become part of the "case file." If the applicant refuses to participate or is uncooperative or unresponsive in the course of the interviews, this fact will be included in the statement and report filed by the chaplain and psychiatrist or medical officer.

E-27. **6.4.3.6.** The investigating officer's report, along with the individual's application, all interviews with chaplains or doctors, evidence received as a result of the investigating officer's hearing, and any other items submitted by the applicant in support of his case will constitute the record. The investigating officer's conclusions and recommended dispositions will be based on the entire record and not merely on the evidence produced at the hearings. A copy of the record will be furnished to the applicant at the time it is forwarded to the commander who appointed the investigating officer, and the applicant will be informed that he has the right to submit a rebuttal to the report within the time prescribed by the Military Service concerned.

E-28. **6.6.** The Secretary of a Military Service may delegate authority to approve applications to the commander exercising general court-martial jurisdiction (or equivalent level command for Reserve organizations) over the applicant. The completed record of a case approved in the field will be forwarded to the Headquarters of the Military Service concerned for appropriate disposition.

E-29. **6.9.** To the extent practicable under the circumstances, during the period applications are being processed and until a decision is made, every effort will be made to assign applicants to duties, which will conflict as little as possible with their asserted beliefs. Unless the Military Service concerned provides otherwise, an applicant shall be required to comply with active duty or transfer orders in effect at the time of his application or subsequently issued and received. During the period applications are being processed, applicants will be expected to conform to the normal requirement of military service and to perform such duties as are assigned. Applicants may be disciplined for violations of the Uniform Code of Military Justice while awaiting action on their applications.

DODD 1300.17 – Accommodation of Religious Practices within the Military Services

E-30. References:

- DOD Directives 1300.17, subject as above, June 18, 1985 (hereby canceled)\
- Public Law 98-525, section 554(d), DOD Authorization Act, 1985, October 19, 1984
- Public Law 100-180, section 508, DOD Authorization Act, 1988-89, December 4, 1987
- Title 10, United States Code, Chapter 47, Uniform Code of Military Justice

E-31. **3.1.** A basic principle of our nation is free exercise of religion. The Department of Defense places a high value on the rights of members of the Armed Forces to observe the tenets of their respective religions. It is DOD policy that requests for accommodation of religious practices should be approved by commanders when accommodation will not have an adverse impact on military readiness, unit cohesion, standards, or discipline.

E-32. **3.2.1.** Worship services, holy days, and Sabbath observances should be accommodated, except when precluded by military necessity.

E-33. **3.2.2.** The Military Department should include religious belief as one factor for consideration when granting separate rations, and permit commanders to authorize individuals to provide their own supplemental food rations in a field or "at sea" environment to accommodate their religious beliefs.

E-34. **3.2.3.** The Military Department should consider religious beliefs as a factor for waiver of immunizations, subject to medical risks to the unit and military.

E-35. **3.2.6.** Religious items or articles not visible or otherwise apparent may be worn with the uniform, provided they shall not interfere with the performance of the member's military duties, as discussed in subparagraph

3.2.7.5. below, or interfere with the proper wearing of any authorized article of the uniform.

E-36. **3.2.7.** Under Public Law 100-180, section 508 (reference (C)), members of the Armed Forces may wear visible items of religious apparel while in uniform, except under circumstances in which an item is not neat and conservative or its wearing shall interfere with the performance of the member's military duties.

E-37. **3.2.7.1.** Under this Directive, "religious apparel" is defined as articles of clothing worn as part of the doctrinal or traditional observance of the religious faith practiced by the member. Hair and grooming practices required or observed by religious groups are not included within the meaning of religious apparel, Jewelry bearing religious inscriptions or otherwise indicating religious affiliation or belief is subject to existing Service uniform regulation just as jewelry that is not of a religious nature.

E-38. **3.2.7.2.** In the context of the wearing of a military uniform, "neat and conservative" items of religious apparel are those that:

E-39. **3.2.7.2.1.** Are discreet, tidy, and not dissonant or showy in style, size, design brightness, or color.

E-40. **3.2.7.2.2.** Do not replace or interfere with the proper wearing of any authorized article of the uniform

E-41. **3.2.7.2.3.** Are not temporarily or permanently affixed or appended to any authorized article of the uniform.

E-42. **3.2.7.3.** The standards in subparagraph 3.2.7.2., above, are intended to serve as a basis for determining a member's entitlement under Public Law 100-80, section 508 (reference ©), to wear religious apparel with the uniform. For example, unless prohibited by subparagraph 3.2.7.6., below, a Jewish yarmulke may be worn with uniform whenever a military cap, hat or other headgear is not prescribed. A yarmulke may also be worn underneath military headgear as long as it does not interfere with the proper wearing, functioning, or appearance of the prescribed headgear.

E-43. **3.2.7.4.** Exceptions to the standards in subparagraph 3.2.7.2., above, and other special accommodations for members of particular religious groups may be granted by the Military Departments under section 4., below.

E-44. **3.2.7.5.** Whether an item of religious apparel interferes with the performance of the member's military duties depends on the characteristics of the items, the circumstances of its intended wear, and the particular nature of the member's duties. Factors in determining if an item of religious apparel interferes with military duties include, but are not limited to, whether the item may.

E-45. **3.2.7.5.1.** Impair the safe and effective operation of weapons, military equipment, or machinery.

E-46. **3.2.7.5.2.** Pose a health or safety hazard to the wearer or others.

E-47. **3.2.7.5.3.** Interfere with the wearing or proper functioning of special or protective clothing or equipment (e.g., helmets, flack jackets, flight suits, camouflaged uniforms, gas masks, wet suits, and crash and rescue equipment).

E-48. **3.2.7.5.4.** Otherwise impair the accomplishment of the military mission.

E-49. **3.2.7.6.** A complete prohibition on the wearing of any visible items of religious apparel may be appropriate under unique circumstances in which the member's duties, the military mission, or the maintenance of discipline require absolute uniformity. For example, members may be prohibited from wearing visible religious apparel while wearing historical or ceremonial uniforms; participating in review formations parades honor or color guards, and similar ceremonial details and functions.

E-50. **3.2.7.7.** The authority to approve the wearing of an item of religious apparel with the uniform, under the guidelines of this paragraph, shall be exercised at the command level specified by each Military Department. Denials of requests to wear religious apparel shall be subject to review oat the Service Headquarters level. Final review shall occur within 30 days following the date of initial denial for cases arising in the United States, and within 60 days for all other cases. Exceptions to these deadlines shall be limited to exigent circumstances, such as extended deployment. Service. Service members shall be obliged to comply with orders prohibiting the wearing of questionable items of religious apparel pending review of such order under regulations issued by the Secretaries of the Military Departments.

E-51. **3.2.7.8.** Notwithstanding paragraphs 3.2.6. and 3.2.7. above, chaplains may wear any required religious apparel or accouterments and the uniforms while conducting worship services and during the performance of rites and rituals distinct to their faith groups.

PROCEDURES 4.

E-52. **4.1.** Under rules prescribed by the Secretary of the Military Department concerned, military commanders should consider the following factors along with any other factors deemed appropriate in determining whether to grant a request for accommodation religious practices addressed in section 3., above.

E-53. **4.1.1.** The importance of military requirements in terms of individual and unit readiness, health and safety, discipline, morale, and cohesion.

E-54. **4.1.2.** The religious importance of the accommodation to the requester.

E-55. **4.1.3.** The cumulative impact of repeated accommodations of a similar nature.

E-56. **4.1.4.** Alternative means available to meet the requested accommodation.

E-57. **4.1.5.** Previous treatment of the same or similar requests, including treatment of similar requests made for other than religious reasons.

E-58. **4.2.** The factors in subsection 4.1. above are intended to promote standard procedure for resolving difficult questions involving accommodation of religious practices. In view of the different mission requirements of each command, individual consideration of specific requests for accommodation is necessary. With the exception of requests involving the wearing of visible items of religious apparel with the uniform, denials of which must be reviewed at the Service Headquarters level, the appropriate level of command for resolution of these issues shall be determined by each of the Military Departments, based on its particular requirements and circumstances.

E-59. **4.3.** When requests for accommodation are not in the best interest of the unit and continued tension between the unit's requirements and the individual's religious belief is apparent, administrative actions should be considered. These actions may include, but are not limited to, assignment, reclassification, or separation. Nothing in this Directive precludes action under the Uniform Code of Military Justice (reference (d) in appropriate circumstances.

USC, TITLE 18, POSSE COMITATUS ACT, SECTION 1385

E-60. The Posse Comitatus Act prohibits the use of the Army and Air Force to enforce civilian law (state or federal) except in cases and under circumstances expressly authorized by the constitution or act of congress. DOD policy extends this prohibition to include the Navy and Marine Corps.

E-61. This act restricts federal military law enforcement activities as follows—

- DOD military personnel do not participate directly—
  - In arrest, search and seizure, or stopping and frisking of persons, or interdiction of vessels, aircraft, or vehicles.
  - In surveillance or pursuit.
  - As informants, undercover agents, or investigators in civilian legal cases or in any other civilian law enforcement activity.
- The Posse Comitatus Act does not apply to —
  - Members of the National Guard when in Title 32 and state active duty (SAD).
  - Members of a Reserve Component when not on active duty or active duty for training.
  - DOD civilians, unless under the direct C2 of a Title 10 active duty officer.
  - Military personnel when off duty and acting solely in a private capacity.
  - Military personnel taking action for the primary purpose of furthering a military or foreign affairs function of the U.S. For example, enforcing military justice, maintaining law and order on military installations, or protecting classified materials.
  - The U.S. Coast Guard.

Appendix F

GUIDE FOR RELIGIOUS AREA/IMPACT ASSESSMENT

As the Army deploys to more areas of the world not familiar to Americans, it becomes increasingly important for chaplains to advise commanders and soldiers on the religious/spiritual aspects of the people they encounter. The term "religious/spiritual" refers to the system of beliefs and practices that give meaning and purpose to people's lives. The Unit Ministry Team (UMT) task is to understand and assess and to remain objective.

This guide gives general parameters to help the UMT fulfill its mission. Not all religious/spiritual systems will contain all of these categories. Rather they are a general matrix that will serve as a starting point to help the UMT in its assessment. The chaplain needs to first consider which of the categories are applicable to the particular society to which he has deployed. This can be done through consulting relevant literature or local subject matter experts. Having determined which categories are relevant, the chaplain can then proceed to assess how the local population understands these categories. The final step would be to covey this knowledge to commanders and soldiers as appropriate. The chaplain's answer to the questions below must address in detail how the chaplain's answer is of issue to the commander as he prepares to conduct operations.

F-1. Holy Day, Rituals and Customs—

- What Holy days and/or festivals may impact military operations?
- What are important religious "do's and don'ts" to observe?
- What are dietary restrictions?
- What are the group's distinctive symbols?
- What are the nature, frequency and traits of worship?
- What are appropriate protocols for issues related to birth/marriage/death?

F-2. Sites and Shrines—

- What are the places of worship, pilgrimage and memorial sites? Why?
- Where are the cemeteries and what is the character of their make-up?
- What are distinctive identifying characteristics (architecture, symbols, etc.)?

F-3. Primary Values—

- What are they willing to die for?

- What subjects incite an emotional response?
- What behaviors does the group reward? What are punished?
- What value is placed on women, children, ancestors, certain animals or objects?

F-4. Leadership—

- Who are the religious leaders (official and unofficial)?
- Do the religious leaders have an impact on the armed forces?
- What political influence do religious leaders have?
- What is the relationship of religious leaders to government officials?
- What do religious leaders wear to symbolize their position?
- How many leaders are there and where are they located?
- What are the leaders titles?
- How are leaders selected and trained?

F-5. Tolerance/Religious Intensity—

- What is the religious commitment in this group?
  - Nominal (in name only)
  - Mild
  - Strong
  - Radical / Fanatical
- How tolerant is the group to other parties?
- How accepting are they of conversion of their members to other groups?
- How are members disciplined?
- Can others join and quit the group easily?
- How are competing groups viewed and received?

F-6. Relationship to Society—

- How does this society relate to the religious group?
  - Stamp out the group
  - Contain the group
  - Assimilate (absorb) the group
  - Share power with the group
  - Promote pluralism with this group
- How is this group viewed?
  - Religious group
  - Secret society
  - Protest movement
  - Political party
- Does the group have a distinct subculture or communal life?
- How does the group seek to influence society?
- How do they use media resources?

F-7. Organization—

- What cell-like groups are present?
- What is the nature of hierarchy within the movement?
- What are the centers of learning?

F-8. Doctrines/Myths—

- What is/are the sources of doctrinal authority?
- What is/are the sources of ethics?
- What are their concepts of justice?
- What are their concepts of reward?
- Who are the heroes and villains? Rivalries (past, present, future)?

F-9. History/Background—

- What larger group is this group related to?
- What makes this group distinctive from the larger group?
- When did this group come into the area?

F-10. References/Sources

Appendix G

CHAPLAIN DETACHMENTS

Chaplain Detachments (CDs) provide flexibility, robustness and specific liaison functions for religious support. CDs represent a supplemental capability for conducting religious support operations and for providing/performing religious support across the full spectrum of military operations. When mobilized, these detachments increase the commander's ability to meet the religious support needs of soldiers.

There are four types of detachments:

- Chaplain Detachment–Alpha (CD-A), General Support (GS)
- Chaplain Detachment–Bravo (CD-B), (GS)
- Chaplain Detachment–Corps (CD-C), Direct Support (DS)
- Chaplain Detachment–Division (CD-D), (DS).

When mobilized, the CDs will normally come from the Reserve Components. It is important to note that these detachments are augmentation resources, not replacements for active component unit ministry teams (UMTs).

CD-A and CD-B UMTs offer maximum religious support flexibility for utilization primarily in major combat operations, but may also be utilized in homeland security and non-major combat operations. If directed by the Chief of Chaplains or FORSCOM, CD-A and CD-B UMTs may coordinate and/or provide religious support to first-responders, units without organic UMTs, DOD civilians, federal agencies, and other personnel as directed during MACA operations, WMD or other HLS incident. However, the CD-C and CD-D UMTs address specific critical religious support operational requirement shortfalls needed for the accomplishment of religious missions in the corps and division area of operations. These detachments will be aligned to support a specific corps or division. The CD-C and CD-D will have a habitual relationship for training and operational missions.

All CDs will rely on the unit to which assigned or attached for all life support and logistical support. Support may also be provided by other appropriate elements within the area of operation. This support includes, but is not limited to: legal, combat health support, finance, personnel and administrative services, field feeding, unit maintenance, and logistical support.

CHAPLAIN DETACHMENT- ALPHA (GENERAL SUPPORT)

G-1. *Mission.* The primary mission of the Chaplain Detachment – Alpha (CD-A) is to provide religious area support across the full spectrum of military operations, and spiritual reconstitution/recollection (SR2) in a Port of Debarkation (POD), Army, Theater, Corps, or equivalent separate task force. The CD-A also provides C2, operational planning and coordination support of Chaplain Detachment – Bravo (CD-B). The CD-A has the following capabilities:

- Planning, supervision, coordination, and provision of religious support for units operating in major ports of debarkation.
- Provide religious support for replacements and Force Reception, Onward movement operations in marshaling areas and ports of debarkation (PODs).
- Plan for and provide operational, specialized religious support and SR2 activities in rest, stand down, and convalescent/recovery areas. This detachment can conduct up to 2 activities per day.
- Plan for and provide area and denominational support for unit concentrations.
- Provide religious, indoctrination, and acclimatization support to incoming personnel, UMT replacements and UMTs during the reception, training, and acclimatization phase of the mobilization.
- Plan and provide religious support activities for units undergoing reconstitution.
- Provide religious support during NEOs.
- Conduct religious support operations in a JTF headquarters during operations other than war.
- Provide technical command and control (C2), supervision, control, and coordination of up to 5 CD-B UMTs.
- Plan and coordinate CD-B religious support operational missions.
- Provide the ASCC, Theater Support Command (TSC) or Corps chaplain with plans and recommendations for integrating specialized religious support and SR2 into contingency plans (COPLANs), operational plans (OPLANs) and operation orders (OPORDs) to enable support to unit/task force personnel.
- Perform other area and denominational religious support missions as required.

G-2. *Personnel—*

- Two chaplains (56A).
- Three chaplain assistants (56M).

G-3. *Assignment.* A CD-A will be assigned to an Army Service Component Command (ASCC), TSC, corps or equivalent separate task force headquarters. The unit will also be assigned as part of the port opening package of major PODs for religious support coverage of port elements. It may also be assigned to a task force or JTF headquarters.

G-4. *Employment.* A CD-A is used to augment religious support capabilities in rear areas from the division rear boundary through the communications zone in units and geographic areas where UMTs are not available or are unable to handle the religious support requirement, particularly areas such as ports of debarkation, marshaling areas, logistic support areas, combat support hospitals and other locations where there is a high concentration of soldiers. CD-A will be deployed as part of port opening packages to conduct religious support operations in major ports for port operational elements without organic UMTs. Additionally, it can be used with Force Provider or the Containerized Chapel to provide religious support to units in transit and to support personnel replacements arriving in theater. A CD-A can perform "spiritual reconstitution/recollection" activities for soldiers and members of other UMTs. If required, the CD-A can divide into two teams to provide expanded area support.

G-5. *Mobility*. The CD-A requires 100% mobility. It must have all required TOE equipment, supplies and personnel when mobilized in order to have 100% operational capability once deployed. This detachment will be required to operate independently throughout the theater, from the port(s) of debarkation (POD) to the division rear boundary while maintaining communication with the UMT of the unit to which assigned/attached. It must be able to provide technical control and coordination of up to 5 CD-Bs throughout the theaters area of operation. This detachment must be able to move all personnel and equipment in a single lift with its authorized organic vehicles.

G-6. Communications Equipment. The CD-A has a requirement to be fielded with the latest technology in both secured voice and digital data communications equipment. The CD-A will be required to communicate digitally and by voice with the unit(s) they are supporting (ASCC, TSC, or Corps or JTF HQs), CD-B UMTs, and back to home station from anywhere on the battlefield. In order for this detachment to meet its minimum mission essential wartime requirements, it must have communications equipment that is compatible and interoperable with the equipment used by the command UMT element of the unit to which assigned and units supported.

CHAPLAIN DETACHMENT-BRAVO (GENERAL SUPPORT)

G-7. *Mission*. The primary mission of the Chaplain Detachment – Bravo (CD-B) is to provide religious support by augmenting the ASCC, TSC, or Corps chaplain sections ability to provide unit and denominational religious support across the full spectrum of military operations at PODs, LSAs, base clusters, and for units without organic religious support assets or those units that deploy under concepts of split-based operations or modularity. It also provides religious support during NEOs, to units undergoing reconstitution, and mass casualties (MASCALs). The CD-B has the following capabilities:

- Provide and perform religious support activities to small units that deploy under the concepts of split-based operations or modularity and to units that do not have organic religious support assets.
- Provide and perform religious support activities for units from other services provided that, the unit to which the CD-B is assigned/attached has Army Support of other Services (ASOS) responsibility and that the unit does not have organic religious support assets.
- Provide and perform comprehensive religious support to family members prior to and during NEOs in forward deployed theatres.

- Provide and perform religious support activities for units undergoing reconstitution.
- Provide religious support to DOD civilians, contractors, and other authorized personnel, as dictated by the commander, that are in the area of operations.
- Provide SR2 when augmenting the CD-A.
- Deploy forward for immediate temporary UMT casualty replacement in divisional maneuver units as required for religious support.
- Provide other area and denominational religious support missions as required.

G-8. *Personnel—*

- One chaplain (56A).
- One chaplain assistant (56M).

G-9. *Assignment.* One CD-B will be assigned to each Corps Support Command (COSCOM) and Corps Support Group (CSG) to augment the chaplain section's ability to provide religious support within the AO. This unit can also be assigned to a TSC, ASG, or BSB.

G-10. *Employment.* Chaplain Detachment-Bravo's are used to augment religious support capabilities throughout the Theatre Army's and Corps area of operations. CD-Bs will normally be employed with a COSCOM or CSG. They can also be employed with, assigned to, or attached to Area Support Groups (ASG) or Base Support Battalions (BSB) to augment the religious support capabilities of those UMTs. Augmented UMTs will employ the CD-B to fill gaps in religious support coverage, specifically to units without organic UMTs within the assigned units area of responsibility. This would also include small units deployed under the concept of split-based operations, specifically at LSAs, and base clusters. This unit can also be utilized to assist Combat Support Hospitals (CSH) UMTs in the event of mass casualties. They can further be pushed down to the Division area of operations to support mass casualty operations when available UMT assets are unable to provide the required religious support. CD-B UMTs will not be employed as permanent UMT casualty replacements in maneuver units, but can be used as immediate temporary (usually 72 hours or less) backfill.

G-11. *Mobility.* The CD-B requires 100% mobility. It must have all required TOE equipment, supplies and personnel when mobilized in order to have 100% operational capability once deployed. This detachment will be required to operate independently throughout the theater, from the PODs to the division rear boundary. At times this detachment will also operate within the division battlespace to provide area and denominational religious support. This detachment must remain prepared to relocate on short notice (3 hours or less). It must also be able to move all personnel and equipment in a single lift with its authorized organic vehicles.

G-12. *Communications Equipment.* The CD-B has a requirement to be fielded with the latest technology in both secured voice and digital data communications equipment. The CD-B will be required to communicate digitally and by voice with the unit(s) they are supporting (ASCC, TSC, Corps, COSCOM, CSG, ASG, and BSB) and CD-A UMTs from anywhere on the battlefield. When providing temporary UMT backfill to maneuver brigades and battalions, the CD-B will be required to communicate with the division and/or maneuver brigade UMT to which they are operationally controlled to. In order for this detachment to meet its minimum mission essential wartime requirements, it must have communications equipment that is compatible with the equipment used by the command UMT element to which assigned and the unit supported.

CHAPLAIN DETACHMENT – CORPS (DIRECT SUPPORT)

G-13. *Mission.* The primary mission of the Chaplain Detachment – Corps (CD-C) is to augment the Corps Chaplain Section across the full spectrum of military operations to provide liaison with the G5/Civil Military Operation Center to facilitate the UMT supported Civil Military Operations activities within the corps on behalf of the corps chaplain. The CD-C is a religious support asset. The CD-C has the following capabilities:

- Plan for and coordinate religious support missions in support of area civil/military operations.
- Provide liaison with government, faith-based NGOs and International Organizations (IOs).
- Coordinate and monitor faith-based NGO and IO requirements and issues.
- Establish links with faith-based NGOs and IOs and with religious leaders of the host nation.
- Plan for and provide operational, specialized religious support and SR2 activities in rest, stand down, and convalescent/recovery areas.
- Provide the Corps Chaplain with plans and recommendations for contingency and reconstitution plans and support to the task force.
- Conduct specialized ministry/SR2 in the corps area at sites selected by using commanders.
- Perform other religious support missions as required.

G-14. *Personnel—*

- One chaplain (56A).
- One chaplain assistant (56M).
  - The chaplain and the chaplain assistant will have a need for specialized training in civil affairs, civil-military operations, and world religions to more effectively perform their mission. They should also have a foundational understanding and be familiar with the operations, missions, and goals of known faith-based NGOs and IOs.
  - The chaplain assistant will be required to have a 2S ASI (Battle Staff Qualified). This skill is required in order for the CD-C to work with the CMOC and the corps chaplain section in developing plans and operations. This skill is also important because it will allow the CD-C to maintain situational awareness and an understanding of the operational environment.

G-15. *Assignment.* The CD-C is assigned to all Heavy and Airborne Corps. They can further be attached or assigned to a JTF HQs or division when the division is operating as an ARFOR or JTF headquarters. The CD-C will habitually train and deploy with the corps to which assigned.

G-16. *Employment.* The CD-C will deploy with and provide habitual augmentation to the Corps chaplain section UMT. The CD-C will act as the liaison between faith-based NGO/IOs and the corps chaplain in an area of operation. The CD-C will be the principle point of contact between the Corps Chaplain and the CMOC concerning indigenous religious issues and religious support impact during Civil Military Operations missions. The CD-C may provide general religious support as the primary mission permits. The CD-C will normally co-locate with the Corps Chaplain section at the corps rear CP and operate out of either the corps chaplain cell or the CMOC. If required the CD-C can be deployed in support of a division that has ARFOR/JTF responsibility in situations where the corps HQs does not deploy. It may also be deployed forward into a division AO to conduct CA/HA and religious support liaison functions.

G-17. *Mobility.* The CD-C requires 100% mobility. It must have all required TOE equipment, supplies and personnel when mobilized in order to have 100% operational capability once deployed. This detachment will be required to operate independently throughout the corps area of operations. At times it will also be required to operate within the division area to conduct support to CMO and religious support liaison functions. This detachment will prepare to be separated from the Corps Chaplain Section for up to 72 hours. It must also be able to move all personnel and equipment in a single lift with its authorized organic vehicles.

G-18. *Communications Equipment.* The CD-C has a requirement to be fielded with the latest technology in both secured voice and digital data communications equipment. The CD-C will be required to communicate digitally and by voice with the corps and division chaplain sections from anywhere within the area of operations. They will also be required to communicate with faith-based NGO/IO representatives, host nation religious leaders, and other governmental organizations operating in the area of operations by voice communications equipment and face-to-face using either military linguists or a machine translation devices The CD-C will also have a CONUS reach-back requirement in order to acquire information on indigenous religions and faith-based NGO/IOs from subject matter experts, databases, and the World Wide Web. In order for this detachment to meet its minimum mission essential wartime requirements, it must have communications equipment that is compatible with the equipment used by the command UMT element to which assigned and other units supported.

CHAPLAIN DETACHMENT –DIVISION (DIRECT SUPPORT)

G-19. *Mission.* The primary mission of the Chaplain Detachment – Division (CD-D) is to provide direct habitual augmentation to the Division Chaplain section. Provides assets required to enable the division chaplain section to conduct 24-hour and split-based religious support planning, coordination, and execution operations across the full spectrum of military operations. The CD-D is capable of the following:

- Provides required assets to the division chaplain section to enable sustained 24-hour and split-based operations between the DREAR and DMAIN.

- Battle track and monitor current division operations and subordinate unit and UMT locations.
- Operates battlefield-automated systems (BAS) to receive and analyze data for recommendations to the division chaplain for use in current operations and future planning of religious support.
- Collects SITREPs, reports, and requests for information (RFIs) from division CP sections/elements, corps chaplain section, brigade and separate battalion UMTs.
- Monitors locations and religious support needs and coordinates religious support for authorized civilians during operations to include: contractors, DA civilians, and interagency representatives operating in the division AO.
- Monitors location of faith-based NGO/IOs operating in the division AO and assesses religious support impact on operations.
- Perform other religious support missions as required.

G-20. *Personnel—*

- One chaplain (56A).
- One chaplain assistant (56M). The chaplain assistant will be required to have a 2S ASI (Battle Staff Qualified). This skill is required to enable the CD-D to be able to battle track and analyze current operations as a means to effectively anticipate religious support needs and develop input into current and future operational plans. It is also important because it will allow the CD-D to maintain situational awareness and an understanding of the operational environment; allowing for more accurate information exchanges to the division chaplain and staff regarding religious support operations.

G-21. *Assignment.* A CD-D will be assigned to each division.

G-22. *Employment.* A critical task for the division chaplain section is 24-hour religious support operations. The CD-D will be aligned to support and train with a specific division in order to develop a habitual relationship and operational knowledge of the division supported. Upon deployment notification of the division, the CD-D that has been assigned to the deploying division will be mobilized to deploy with that division. The CD-D will join the division chaplain section during the pre-deployment phase of the operation. This will permit the division chaplain section to conduct 24-hour and/or split-based religious support operations throughout the course of the mission and across the full spectrum of military operations. The CD-D will normally operate out of the DMAIN along with the division chaplain and NCOIC. While the CD-D can be used to perform general religious support functions, it is important to remember that the primary mission of the CD-D is to assist in performing 24-hour operations.

G-23. *Mobility.* The CD-D requires 100% mobility. It must have all required TOE equipment, supplies and personnel when mobilized in order to have 100% operational capability once deployed. This detachment will be required to have the same mobility as the division it supports. It must also be able to move all personnel and equipment in a single lift with its authorized organic vehicles.

G-24. *Communications Equipment.* The CD-D has a requirement to be fielded with the latest technology in both secured voice and digital data communications equipment. The CD-D will be required to communicate digitally and by voice with the division UMT, deputy division UMT, BDE and BN UMTs throughout the division AO. In order for this detachment to meet its minimum mission essential wartime requirements, it must have communications equipment compatible with the equipment used by the division chaplain section to which assigned and other units supported.

GLOSSARY

AA	assembly area
AAR	After Action Review
ABCA	American, British, Canadian, and Australian
AC	Active component
ACR	Armored Cavalry Regiment
ADA	air defense artillery
AGR	Active Guard Reserve
ALOC	Administrative/Logistics Operations Center
AO	area of operations
AOB	advanced operational bases
AOR	area of responsibility
AMC	Army Materiel Command
AMEDD	Army Medical Department
APOE	aerial port of embarkation
AR	Army regulation
ARFOR	Army forces
ARNG	Army National Guard
ARSOA	Army Special Operations Aviation
ARSOF	Army Special Operations Forces
ARTEP	Army Training and Evaluation Program
ASCC	Army Service Component Command
ASG	area support group
AT	annual training
AUTL	Army Universal Task List
AXP	ambulance exchange point
BAS	battalion aid station
BASOPS	base operations
BCOC	Base Cluster Operations Center
BCTP	Battle Command Training Program
BDE	brigade
BDOC	Base Defense Operations Center
BII	Basic issue items
BMO	battalion maintenance officer
BN	battalion
BOS	battlefield operating system
BSA	brigade support area
BSB	base support battalion

BUB	battle update briefings
C2	command and control
CCC (C3)	chaplain career course
C4ISR	C4 Intelligence, surveillance and Reconnaissance
CA	chaplain assistant
CA	Civil Affairs
CCH	Chief of Chaplains
CCP	Casualty Collection Point
CD	Chaplain Detachment
CD-A	Chaplain Detachment-Alpha (General Support)
CD-B	Chaplain Detachment-Bravo (General Support)
CD-C	Chaplain Detachment-Corps (Direct Support)
CD-D	Chaplain Detachment-Division (Direct Support)
CED	Critical Event Debriefing
CEOI	Communication electronic operating instructions
CFLC	Chaplain Family Life Center
CFP	Contingency Force Pool
CH	Chaplain
CHOBC	Chaplain Officer Basic Course
CISM	Critical Incident Stress Management
CJTF	combined-joint task force
CJTF	commander joint task force
CMM	Combat Medical Ministry Course
CMO	civil-military operations/officer
CMOC	Civil Military Operations Center
CMRP	Command Master Religious Plan
CMTC	Combat Maneuver Training Center
COA	course of action
CofS	Chief of Staff
COMMZ	communications zone
CONUS	Continental United States
CONUSA	Continental United States Army
COPLAN	contingency plan
COSCOM	Corps Support Command
COSR	combat operations stress response
CP	command post
CPE	Clinical Pastoral Education
CRC	Continental United States Replacement Center
CRP	Commander's Religious Program
CS	combat support
CSA	corps support area

CSB	Corps Support Battalion
CSG	Corps Support Group
CSM	Command Sergeant Major
CSS	combat service support
CSSCS	Combat Service Support Control System
CSU	Chaplain Support Unit
CT	Counter Terrorism
CTC	Combat Training Center
CTCP	Combat Trains Command Post
CZ	combat zone
DA	Department of the Army
DA	Direct Action
DACH	U.S. Army Chief of Chaplains
DASG	U.S. Army Surgeon General
DCO	Defense Coordinating Office
DCSOPS	Deputy Chief of Staff for Military Operations and Plans
DFG	Distinctive Faith Group
DFGL	Distinctive Faith Group Leaders
DISCOM	Division Support Command
DIV (IT)	Institutional Training Divisions
DIVARTY	division artillery
DLA	Defense Logistics Agency
DNVT	digital nonsecure voice telephone
DOD	Department of Defense
DOS	Days of Supply
DS	direct support
DSA	division support area
DSCP	Defense Supply Center Philadelphia
DSL	denominational service leader
DSO	domestic support operations
DTG	date time group
DTLOMS	doctrine, training, leader development, organization material, and soldier support
EAC	echelons above corps
EAD	echelons above division
EMM	Emergency Medical Ministry Course
EPW	enemy prisoner of war
FA	field artillery
FAC	family assistance center
FEMA	Federal Emergency Management Agency
FHA	foreign humanitarian assistance
FID	Foreign Internal Defense

FLOT	forward line of own troops
FORSCOM	Forces Command
FM	frequency modulation
FRAGO	fragmentary order
FRG	Family Readiness Group
FSCOORD	fire support coordinator
FSMC	forward support medical company
FTCP	Field Trains Command Post
FTX	field training exercises
GCI	Group Crisis Intervention
GS	general support
GSU	Garrison Support Unit
HA	humanitarian assistance
HHC	headquarters and headquarters company
HHT	headquarters and headquarters troop
HN	host nation
HQDA	Headquarters, Department of the Army
HSSN	home station support node
HUMINT	human intelligence
IET	initial entry training
IMA	Individual Mobilization Augmentee
IMT	Installation Ministry Team
IO	Information Operations
IPB	intelligence preparation of the battlefield
IRR	Individual Ready Reserve
ISB	intermediate staging base
JAG	Judge Advocate General
JAS	Jump Aid Station
JFC	Joint Force Commander
JFCOM	Joint Forces Command
JFLC	Joint Forces Land Component
JFLCC	Joint Forces Land Component Command
JFSOCC	Joint Force Special Operations Component Commander
JSOTF	Joint Special Operations Task Force
JOA	joint operational area
JOPES	Joint Operation Planning and Execution System
JRTC	Joint Readiness Training Center
JTF	joint task force
JSOA	Joint Special Operations Area
KIA	killed in action
LD	line of departure

LDRSHIP	loyalty, duty, respect, selfless service, honor, integrity, and personal courage
LFA	Lead Federal Agency
LL	Lessons Learned
LOGPAC	Logistics Package
LRP	logistics release point
MACA	military assistance to civil authorities
MACDIS	military assistance to civil disturbances
MACOM	Major Army Command
MASCAL	mass casualty
MDMP	Military Decision Making Process
MEDCOM	U.S. Army Medical Command
METL	mission essential task list
METT-TC	mission, enemy, terrain and weather, troops and support available, time available, and civil considerations
MIA	Missing in Action
MOB	Mobilization
MOBEX	mobilization exercises
MOOTW	military operations other than war
MOPP	mission oriented protective posture
MOS	military occupational specialty
MRI	Medical Reengineering Initiative
MRE	meal ready to eat
MS	mobilization station
MSC	Major Subordinate Command
MSCA	military support to civilian authorities
MSE	mobile subscriber equipment
MSR	main supply route
MTF	Medical Treatment Facility
MTOE	Modified Table of Organization and Equipment
NATO	North Atlantic Treaty Organization
NCO	noncommissioned officer
NCOES	Noncommissioned Officer Education System
NCOIC	noncommissioned officer in charge
NDMS	National Disaster Medical System
NEO	noncombatant evacuation operation
NG	National Guard
NGO	non-governmental organization
NMS	National Military Strategy
NOK	next of kin
NOVA	National Organization for Victim Assistance
NSN	national stock number

OCCH	Office of the Chief of Chaplains
OCOKA	Observation and fields of fire, cover and concealment, obstacles, Key terrain, and avenues of approach
OCONUS	Outside the Continental United States
ODA	Operational Detachment Alpha
ODSS	Offensive, Defensive, Stability and Support
OES	Officer Education System
OOTW	operations other than war
OPCON	operational control
OPLAN	operations plan
OPORD	operations order
OPSEC	operations security
OPTEMPO	operation tempo
OVM	operator's vehicle material
PAO	Public Affairs Office/Officer
PBO	Property Book Officer
PCI	precombat inspection
PCS	permanent change of station
PERSCOM	Personnel Command
PHA	Personnel Holding Area
PIES	proximity, immediacy, expectancy, simplicity
PMCS	preventive maintenance checks and services
POD	port of debarkation
POE	port of embarkation
POM	preparation of overseas movement
POTF	PSYOP Task Force
POTG	PSYOP Task Group
POW	prisoner of war
PROFIS	Professional Officer Filler System
PSYOP	psychological operations
RAA	Religious Area Analysis
RAOC	Rear Area Operation Center
RC	Reserve component
RCP	Rear Command Post
RIP	Ranger Indoctrination Program
RFI	request for information
RMST	Religious Ministry Support Team
ROE	rules of engagement
ROP	Ranger Orientation Program
RPP	Religious Preference Profile
RS	religious support
RSC	Reserve Support Command

RSE	religious support estimate
RSOI	reception, staging, onward movement, and integration
RSP	religious support plan
RTF	Reconstitution Task Force
RTOC	Rear Tactical Operation Center
S1/G1	Adjutant/Personnel Officer
S2/G2	Intelligence Officer
S3/G3	Operations and Training Officer
S4/G4	Logistics Officer
S5/G5	Civil Military Operations officer
S6/G6	Signal operations officer
SACRED	spirituality, accountability, compassion, religious leadership, excellence, and diversity
SAD	state active duty
SEAD	Suppression of Enemy Air Defenses
SF	Special Forces
SFOB	Special Forces Operational Base
SFSC	Special Forces Staff Course
SIDPERS	Standard Installation/Division Personnel System
SIGO	signal officer
SITREP	situation report
SJA	Staff Judge Advocate
SMART-B	Special Medical Augmentation Response Team-Burn
SMART-CB	Special Medical Augmentation Response Team-Chemical/Biological
SMART-HS	Special Medical Augmentation Response Team-Health Systems Assessments and Assistance
SMART-MC3T	Special Medical Augmentation Response Team- Medical Command, Control, Communications, Telemedicine
SMART-PC	Special Medical Augmentation Response Team-Pastoral Care
SMART-PM	Special Medical Augmentation Response Team-Preventive Medicine/Disease Surveillance
SMART-SM	Special Medical Augmentation Response Team-Stress Management
SMART-TCC	Special Medical Augmentation Response Team-Trauma/Critical Care
SMART-V	Special Medical Augmentation Response Team-Veterinary
SO	special operations
SOCOM	Special Operations Command
SOF	Special Operation Forces
SOFA	status of forces agreement
SOFCOM	United States Special Operations and Forces Command
SOI	special operating instruction
SOP	standard operating procedures
SPOE	seaport of embarkation

SR	Special Reconnaissance
SR	supply route
SR2	spiritual reconstitution/recollection
SRP	Soldier Readiness Program
SSC	small-scale contingency
SSCO	small-scale contingency operations
STARC	State Area Command, ARNG
TAA	tactical assembly area
TACON	tactical control
TDA	Table of Distribution and Allowances
TF	task force
TF A/L	task force (Administrative/Logistics)
TOA	transfer of authority
TOC	tactical operations center
TSC	Theater Support Command
TSOP	tactical standard operating procedure
TTP	tactics, techniques and procedures
UCMJ	uniformed code of military justice
UMCP	Unit Maintenance Collection Point
UMT	Unit Ministry Team
USACAPOC (A)	United States Army Civil Affairs and Psychological Operations Command (Airborne)
USAR	U.S. Army Reserve
USARC	United States Army Reserve Command
USASOC	United States Army Special Operations Command
USC	United States Code
USDB	United States Disciplinary Barracks
USSOCOM	United States Special Operations Command
UW	unconventional warfare
WIA	wounded in action
WMD	weapons of mass destruction
XO	executive officer

Bibliography

DOCUMENTS NEEDED

These documents should be available to the intended users of this publication.
Except where otherwise indicated, the following DA forms are available on the Army Electronic Library (AEL) CD-ROM (EM 0001) and the USAPA web site (www.usapa.army.mil); DD forms are available from the OSD web site http://www.dior.whs.mil/icdhome/ddeforms.htm

DA Form 7462, Religious Support Matrix, September 2002.

DD Form 1610, Request and Authorization for TDY Travel of DOD Personnel, January 2001.

JOINT AND MULTISERVICE PUBLICATIONS

Most joint publications are online at http://www.dtic.mil/doctrine/

DOD Directive 1300.6, *Conscientious Objectors*, 20 August 1971.

DOD Directive 1300.17, *Accommodation of Religious Practices Within the Military Services*, 3 February 1988.

DOD Directive 1332.14, *Enlisted Administrative Separations*, 21 December 1993.

JP 1-0, *Doctrine for Personnel Support to Joint Operations*, 19 November 1998.

JP 1-05, *Religious Ministry Support for Joint Operations*, 26 August 1996.

JP 3-0, *Doctrine for Joint Operations*, 10 September 2001.

JP 3-05, *Doctrine for Joint Special Operations*, 17 April 1998.

JP 3-07.3, *Joint Tactics, Techniques, and Procedures for Joint Special Operations Task Force Operations*, 19 December 2001.

JP 3-08, *Interagency Coordination During Joint Operations*, Volumes I and II, 9 October 1996.

JP 3-10, *Joint Doctrine for Rear Area Operations*, 28 May 1996.

JP 5-00.2, *Joint Task Force (JTF) Planning Guidance and Procedures*, 13 January 1999.

ARMY PUBLICATIONS

Most Army doctrinal publications are online at http:www.adtdl.army.mil/atdls.html or at http://www.lia.army.mil/dpmsh/pubs_rev_dt.html

AR 165-1, *Chaplain Activities in the United States Army*, 27 February 1998.

AR 380-5, *Department of the Army Information Security Program*, 29 September 2000.

AR 600-20, *Army Command Policy*, 13 May 2002.

AR 680-29, *Military Personnel – Organization and Type of Transactions Codes*, 1 March 1989.

DA PAM 165-3, *Chaplain Training Strategy*, 1 September 1998.

FM 1 (100-1), *The Army*, 14 June 2001.

FM 1-111, *Aviation Brigades*, 27 October 1997.

FM 3-0 (100-5), *Operations*, 14 June 2001.

FM 3-90 (100-40), *Tactics*, 4 July 2001.

FM 3-90.3 (71-3), *The Mounted Brigade Combat Team*, 1 November 2001.

FM 7-0 (25-100), *Train the Force*, 22 October 2002.

FM 7-20, *The Infantry Battalion*, 6 April 1992.

FM 7-30, *The Infantry Brigade*, 3 October 1995.

FM 8-51, *Combat Stress Control in a Theater of Operations Tactics, Techniques and Procedures*, 29 September 1994.

FM 16-1, *Religious Support*, 26 May 1995.

FM 22-51, *Leaders' Manual for Combat Stress Control*, 29 September 1994.

FM 22-100, *Army Leadership*, 31 August 1999.

FM 25-101, *Battle Focused Training*, 30 September 1990.

FM 55-30, *Army Motor Transport Units and Operations*, 27 June 1997.

FM 71-2, *The Tank and Mechanized Infantry Battalion Task Force*, 27 September 1988.

FM 71-100, *Division Operations*, 28 August 1996.

FM 100-6, *Information Operations*, 27 August 1996.

FM 100-7, *Decisive Force: The Army in Theater Operations*, 31 May 1995.

FM 100-8, *The Army in Multinational Operations*, 24 November 1997.

FM 100-9, *Reconstitution*, 13 January 1992.

FM 100-10-2, *Contracting Support on the Battlefield*, 4 August 1999.

FM 100-11, *Force Integration*, 15 January 1998.

FM 100-15, *Corps Operations*, 29 October 1996.

FM 100-17-3, *Reception, Staging, Onward Movement, and Integration*, 17 March 1999.

FM 100-19, *Domestic Support Operations*, 1 July 1993.

FM 100-20, *Military Operations in Low Intensity Conflict*, 5 December 1990.

FM 100-21, *Contractors on the Battlefield*, 26 March 2000.

FM 100-23, *Peace Operations*, 30 December 1994.

FM 101-5, *Staff Organization and Operations*, 31 May 1997.

FM 101-5-2, U.S. Army Report and Message Formats, 29 June 1999

OTHER PUBLICATIONS

Army Vision 2010

Chaplain Assistant Concept Paper, 2000

Chaplain Support Force XXI White Paper

Chief of Chaplains MSCA Concept Paper, September 2000

The Chief of Chaplain's UMT Crisis Assets Handbook, October 2000

Concept for Future Joint Operations, May 1997

Joint Vision 2010

Joint Vision 2020 –http://www.dtic.mil/jv2020

Military Rules of Evidence 503 UCMJ

National Military Strategy

A National Security Strategy for a New Century, May 1997

The Quadrennial Defense Review Report, September 2001.

Public Law 98-525

Public Law 100-180

Textbook of Military Medicine: War Psychiatry, Part I, Warfare, Weaponry and the Casualty Office of the Surgeon General of the United States Army, Falls Church, VA, 1995.

United States Code, Title 10, Armed Forces http://uscode.house.gov/title_10.htm

United States Code, Title 18, Crimes and Criminal Procedures: http://uscode.house.gov/title_18.htm

U.S. Army Chaplaincy Strategic Plan FY 2000-FY 2005

Field Manual
No. 4-02.51 (8-51)

Headquarters
Department of the Army
Washington, DC, 6 July 2006

COMBAT AND OPERATIONAL STRESS CONTROL

Contents

Distribution Restriction: Approved for public release; distribution is unlimited.

***This publication supersedes FM 8-51, 29 September 1994.**

Figures

Tables

Preface

This publication outlines the functions and operations of each combat and operational stress control (COSC) element within an area of operations (AO). This field manual (FM) establishes Army doctrine and provides guidance for conducting COSC support for combat, stability, and reconstruction operations from brigade to theater level. The information provided in this publication will assist commanders and their staffs to operate efficiently at all levels of command and throughout the operational continuum. It may be used by medical planners to supplement FM 4-02, FM 8-42 (4-02.42), and FM 8-55 (4-02.55). Users of this publication should be familiar with Army Regulation (AR) 40-216 and with FM 4-02, FM 4-02.6, FM 4-02.10, FM 4-02.21, FM 4-02.24, FM 6-22.5, FM 8-10-6, FM 8-10-14, and FM 22-51.

This manual is in consonance with FM 7-15, Army Universal Task List (AUTL) and support the following Army tactical task (ART) provided below. Commanders should use the AUTL as a cross-reference for tactical tasks. The AUTL provides a standard doctrinal foundation and catalogue of the Army's tactical collective tasks.

AUTL ART	
ART 6.5.1	Provide Combat Casualty Care
ART 6.5.1.5	Provide Mental Health/Neuropsychiatric Treatment
ART 6.5.4	Provide Casualty Prevention
ART 6.5.4.5	Provide Combat Operational Stress Control Prevention

The staffing and organization structure presented in this publication reflects those established in the base tables of organization and equipment (base TOE) and are current as of the publication print date. Such staffing is subject to change to comply with manpower requirements criteria outlined in AR 71-32. Those requirements criteria are also subject to change if the modification table of organization and equipment (MTOE) is significantly altered.

Users of this publication are encouraged to submit comments and recommendations to improve the publication. Comments should include the page, paragraph, and line(s) of the text where the change is recommended. The proponent of this publication is the United States (US) Army Medical Department Center and School (USAMEDDC&S). Comments and recommendations should be forwarded in a letter format directly to the **Commander, USAMEDDC&S, ATTN: MCCS-FCD-L, 1400 East Grayson Street, Fort Sam Houston, Texas 78234-5052**, or at e-mail address: Medicaldoctrine@amedd.army.mil.

This FM applies to the Active Army, the Army National Guard (ARNG)/Army National Guard of the United States (ARNGUS), and the United States Army Reserve (USAR) unless otherwise stated.

Unless this publication states otherwise, masculine nouns and pronouns do not refer exclusively to men.

Use of trade or brand names in this publication is for illustrative purposes only and does not imply endorsement by the Department of Defense (DOD). The Army Medical Department (AMEDD) is in a transitional phase with terminology. This manual uses the most current terminology; however, other FM 4-02-series and FM 8-series may use the older terminology. Changes in terminology are a result of adopting the terminology currently used in the joint and/or North Atlantic Treaty Organization (NATO) and American, British, Canadian, and Australian (ABCA) Armies publication arenas. Therefore, the following terms are synonymous and the current terms are listed first, to include—

- Force health protection (FHP), health service support (HSS), and combat health support (CHS).
- Medical logistics (MEDLOG), health service logistics (HSL), and combat health logistics (CHL).
- Levels of care, echelons of care, and roles of care.

- Combat and operational stress control (COSC) and combat stress control (CSC).
- Behavioral health (BH) and mental health (MH).

Additionally, please note that the term "battle fatigue (BF)" that is used in AR 40-216 is being replaced with the term "combat stress reaction (CSR)." This change is due to DOD Directive (DODD) 6490.5, which specified that all Military Services use the term CSR for the purpose of joint interoperability. A DOD (Health Affairs) working group with the Services later added the term "operational stress reaction (OSR)" to further characterize stress reactions experienced by Soldiers. The only difference between a CSR and an OSR is that it takes place in a combat environment. To reduce confusion from this change in terminology, this manual will use combat and operational stress reaction (COSR) as opposed to "BF". The COSR (battle fatigue) "casualties" are Soldiers experiencing a stress reaction in combat or operational environment.

Introduction

In our own Soldiers and in the enemy combatants, control of stress is often the decisive difference between victory and defeat across the operational continuum. Battles and wars are won more by controlling the will to fight than by killing all of the enemy combatants. Uncontrolled combat stress causes erratic or harmful behaviors, impair mission performance, and may result in disaster and defeat of COSC preventive measures. The COSC preventive measures are aimed at minimizing maladaptive stress reactions while promoting adaptive stress reactions, such as loyalty, selflessness, and acts of bravery.

This manual provides doctrinal guidance for controlling excessive stress in combat and other operational environments. It identifies command and leadership responsibilities for COSC. It identifies COSC consultation, training, and education assistance available for units. This manual provides definitive guidance to BH personnel and CSC units for their COSC mission and for management of COSR and other behavioral disordered patients (BDPs). It identifies the requirements for COSC consultation, planning, coordination, rehearsal, and implementation of the COSC plan contained in the FHP annex of the operation order (OPORD).

Many stressors in a combat situation are due to deliberate enemy actions aimed at killing, wounding, or demoralizing our Soldiers and our allies. Other stressors are due to the operational environment. Some of these stressors can be avoided or counteracted by wise command actions. Still other stressors are due to our own calculated or miscalculated choice, accepted in order to exert greater stress on the enemy. Sound leadership works to keep stressors within tolerable limits and prepares the troops mentally and physically to endure them. Some of the most potent stressors can be due to personal organizational problems in the unit or on the home front. These, too, must be identified and, when possible, corrected or controlled. Unit needs assessments (UNAs) can help BH providers identify specific stressor in a unit and develop interventions to help unit personnel cope.

This manual identifies MH sections and medical units, CSC, that provides COSC support to units. It identifies the requirements for COSC assistance, as a unit transitions through the different phases of a deployment. Chaplains, commander, leader and COSC Soldier mentors provide assistance with after-action review (AAR) and traumatic events management (TEM).

Chapter 1

Combat and Operational Stress Control

SECTION I — INTERVENTION AND CONTROL FOR THE COMBAT AND OPERATIONAL STRESS THREAT

STRESS CONTROL

1-1. Combat stress includes all the physiological and emotional stresses encountered as a direct result of the dangers and mission demands of combat, see AR 40-216. Combat and operational stress control in the Army may be defined as programs developed and actions taken by military leadership to prevent, identify, and manage adverse COSRs in units. This program optimizes mission performance; conserves the fighting strength; and prevents or minimizes adverse effects of COSR on Soldiers and their physical, psychological, intellectual, and social health. Its goal is to return Soldiers to duty expeditiously. According to DODD 6490.2, COSC activities include routine screening of individuals when recruited; continued surveillance throughout military service, especially before, during, and after deployment; continual assessment and consultation with medical and other personnel from garrison to the battlefield.

1-2. Combat and operational stress control is the commander's responsibility at all levels. The commander is assisted with his responsibility for COSC by his staff, unit leaders, unit chaplain (Appendix A), and organic medical personnel. The commander may also receive assistance from organic COSC personnel at brigade and above, and from corps and above medical company/detachment CSC BH personnel. The key concern to combat commanders is to maximize the return-to-duty (RTD) rate of Soldiers who are temporarily impaired or incapacitated with stress-related conditions or diagnosed behavioral disorders.

1-3. The purpose of COSC is to promote Soldier and unit readiness by—

- Enhancing adaptive stress reactions.
- Preventing maladaptive stress reactions.
- Assisting Soldiers with controlling COSRs.
- Assisting Soldiers with behavioral disorders.

Note. The word control is used with combat and operational stress, rather than the word management, to emphasize the active steps that leaders, supporting BH personnel, and individual Soldiers must take to keep stress within an acceptable range.

COMBAT AND OPERATIONAL STRESS THREAT

1-4. Many stressors in a combat situation are due to deliberate enemy actions aimed at killing, wounding, or demoralizing our Soldiers and our allies. Other stressors are due to the natural environment, such as intense heat or cold, humidity, or poor air quality. Still others are due to leaders' own calculated or miscalculated choices (for example, decisions about unit strength, maneuver, the time of the attack, and plans for medical and logistical support). Sound leadership works to keep operational stressors within tolerable limits and prepares troops mentally and physically to endure them. In some cases however, excessive stress can affect both leaders' and Soldiers' decision-making and judgment, resulting in missed opportunities, or worse, in high casualties and/or failure to complete the mission. Finally, some of the most potent stressors are interpersonal in nature and can be due to conflict in the unit or on the home front. Extreme reactions to such stressors may involve harm to self (as in the hypothetical case of a Soldier that becomes suicidal on discovering that his wife wants a divorce) or to others (as in the case of a Soldier that impulsively fires a weapon at his noncommissioned officer (NCO) out of rage over perceived unfairness). These stressors must be identified and when possible, corrected or controlled. For more information on the control of combat/operational stressors and for details about specific leader actions to control stress, see FM 22-51. Also see FM 4-02 and FM 4-02.17 for additional information on the general and medical threat.

EFFECT OF STRESS

1-5. Focused stress is vital to survival and mission accomplishment. However, stress that is too intense, or prolonged, results in COSR that impairs their ability to function effectively. Some stressors contribute to misconduct that requires disciplinary action and may take the Soldier from duty for legal action and incarceration. In a broader context stress may cause battle and nonbattle injuries through inattention, clumsiness, and reckless behavior. These resultant injuries can include equipment losses and friendly fire incidents. Stress may increase disease rates by disrupting hygiene and protective measures, and impairing the body's immune defenses. Stress may progress to BH disorders, or suicidal behaviors and/or homicidal behaviors. Excessive stress in combat contributes to lapses in operational and tactical judgment and to missed opportunities that could increase the numbers of Soldiers injured over time.

MODEL FOR COMBAT AND OPERATIONAL STRESS CONTROL INTERVENTIONS

1-6. Combat and operational stress control assessments are performed during all phases of combat operations, stability and reconstruction operations, and support operations so the term COSC may refer to any of these different types of Army operations. The COSC assessments are performed at unit and individual level. They consider the range of variables according to a model, which recognizes that biological, psychological, and social factors influence each other. The COSC assessment reviews the interaction systematically to a depth appropriate to the need. The assessment identifies which variables can be modified to improve coping or outcome. Based on these assessments COSC personnel recommend courses of action (COA) to the commander. They identify and initiate COSC interventions to improve unit effectiveness and Soldier efficiency and well-being.

1-7. Provided in Figure 1-1 is a conceptual model of stress, its mitigating and aggravating factors and potential outcomes on Soldiers and families. This model can be helpful when designing COSC interventions to improve short-term and long-term outcomes.

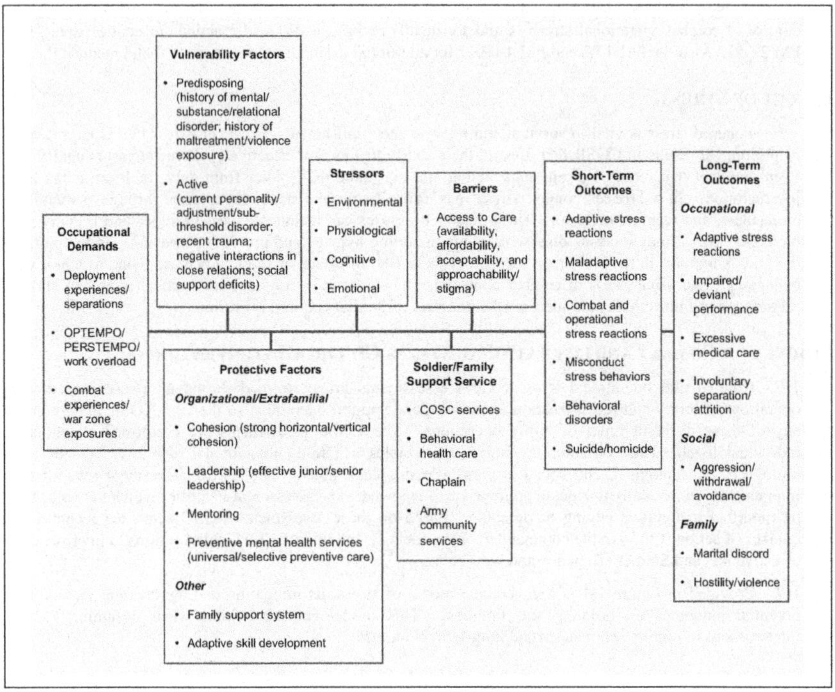

Figure 1-1. Model of stress and its potential Soldier and family outcomes

MENTAL AND PHYSICAL STRESSORS

1-8. A rough distinction between a mental and physical stressor can be made—

- A mental stressor is one in which the brain receives information about a given threat or demand, but this information results in only indirect physical impact on the body. Instead, its primary effect is to place demands on and evoke reactions from the perceptual, cognitive and/or emotional systems of the brain (such as information overload, perceived lack of control, or grief-producing losses).
- A physical stressor has a direct, potentially harmful effect on the body. These stressors may be external environmental conditions (such as temperature) or the internal physiologic demands required by or placed upon the human body (such as the need for hydration, or an immune response to a viral infection).

TYPES OF MENTAL AND PHYSICAL STRESSORS

1-9. Figure 1-2 provides examples for the two types of physical stressors (environmental and physiological) and the two types of mental stressors (cognitive and emotional). Also, as recognized in the COSC intervention model (Figure 1-1) physical stressors cause mental stressors when they result in discomfort, distraction, and threat of harm, as well as when they directly impair brain functions. Mental stressors can lead to adaptive or maladaptive stress behaviors that decrease or increase the exposure to physical stressors.

PHYSICAL STRESSORS	MENTAL STRESSORS
ENVIRONMENTAL	**COGNITIVE**
Heat, cold, wetness, dust Vibration, noise, blast Noxious odors (fumes, poisons, chemicals) Directed-energy weapons/devices Ionizing radiation Infectious agents Physical work Poor visibility (bright lights, darkness, haze) Difficult or arduous terrain High altitude	Information (too much or too little) Sensory overload or deprivation Ambiguity, uncertainty, unpredictability Time pressure or waiting Difficult decision (rules of engagement) Organizational dynamics and changes Hard choices versus no choice Recognition of impaired functioning Working beyond skill level Previous failures
PHYSIOLOGICAL	**EMOTIONAL**
Sleep deprivation Dehydration Malnutrition Poor hygiene Muscular and aerobic fatigue Overuse or underuse of muscles Impaired immune system Illness or injury Sexual frustration Substance use (smoking, caffeine, alcohol) Obesity Poor physical condition	Being new in unit, isolated, lonely Fear and anxiety-producing threats (of death, injury, failure, or loss) Grief-producing losses (bereavement) Resentment, anger, and rage-producing frustration and guilt Inactivity producing boredom Conflicting/divided motives and loyalties Spiritual confrontation or temptation causing loss of faith Interpersonal conflict (unit, buddy) Home-front worries, homesickness Loss of privacy Victimization/harassment Exposure to combat/dead bodies Having to kill

Figure 1-2. Examples of combat and operational stressors

STRESS BEHAVIORS IN COMBAT AND OTHER OPERATIONS

1-10. Combat and operational stress behavior is the generic term that is used for the full spectrum of combat and operational stress behaviors. It covers the range of reactions, from adaptive to maladaptive. Figure 1-3 provides a listing of typical adaptive and maladaptive stress reactions.

ADAPTIVE STRESS REACTIONS

1-11. Stressors, when combines with effective leadership and good peer relationships may lead to adaptive stress reactions which enhance individual and unit performance. Examples of adaptive stress reactions include—

- The strong personal trust, loyalty, and cohesiveness (called horizontal bonding), which develops among peers in a small military unit.
- Personal trust, loyalty, and cohesiveness that develops between leaders and subordinates (called vertical bonding).
- Esprit de corps is also defined as a feeling of identification and membership in the larger, enduring unit with its history and ideas. This may include the unit (such as battalion, brigade

combat team [BCT], regiment, or division) the branch (such as infantry, artillery, or military police [MP]) and beyond the branch to the US Army level.

- Unit cohesion is the binding force that keeps Soldiers together and performing the mission in spite of danger and death. Cohesion is a result of Soldiers knowing and trusting their peers and leaders and understanding their dependency on one another. It is achieved through personal bonding and a strong sense of responsibility toward the unit and its members. The ultimate adaptive stress reactions are acts of extreme courage and almost unbelievable strength. They may even involve deliberate self-sacrifice.

Maladaptive Stress Reactions

1-12. Combat and operational stress reaction and misconduct stress behaviors comprise the maladaptive stress reactions.

Combat and Operational Stress Reaction

1-13. The Army uses the DOD-approved term/acronym COSR in official medical reports. This term can be applied to any stress reaction in the military unit environment. Many reactions look like symptoms of mental illness (such as panic, extreme anxiety, depression, hallucinations), but they are only transient reactions to the traumatic stress of combat and the cumulative stresses of military operations. Some individuals may have behavioral disorders that existed prior to deployment or disorders that were first present during deployment, and need BH intervention beyond the interventions for COSR.

Misconduct Stress Behaviors

1-14. Examples of misconduct stress behaviors are listed in Figure 1-3. These range from minor breaches of unit orders or regulations to serious violations of the Uniform Code of Military Justice (UCMJ) and the Law of Land Warfare. Misconduct stress behaviors are most likely to occur in poorly trained undisciplined Soldiers, however good and heroic, under extreme combat stress may also engage in misconduct. Generally, misconduct stress behaviors—

- Range from minor breaches of unit orders or regulations to serious violations of the UCMJ and the Law of Land Warfare.
- May also become a major problem for highly cohesive and proud units. Such units may come to consider themselves entitled to special privileges and, as a result, some members may relieve tension unlawfully when they stand-down from their military operations. They may lapse into illegal revenge when a unit member is lost in combat. Stress control measures and sound leadership can prevent such misconduct stress behaviors, but once serious misconduct has occurred, Soldiers must be punished to prevent further erosion of discipline. Combat stress, even with heroic combat performance, cannot justify criminal misconduct.

Overlapping Stress Behaviors

1-15. The distinctions among adaptive stress reactions, misconduct stress behaviors, and COSRs are not always clear. Indeed, the three categories of combat and operational stress behaviors may overlap. Soldiers with COSR may show misconduct stress behaviors and vice versa. Soldiers with adaptive stress reactions may also suffer from COSR. Finally, excellent combat Soldiers that have exhibited bravery and acts of heroism may also commit misconduct stress behaviors.

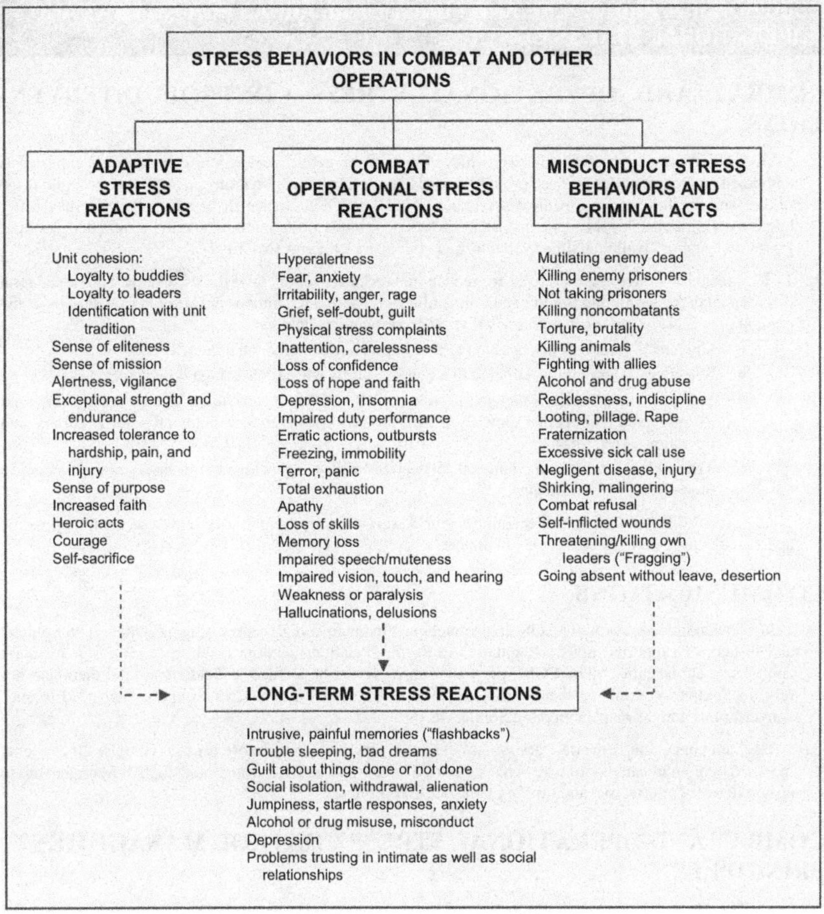

Figure 1-3. Stress behaviors in combat and other operations

1-16. It is common for stress reactions to persist or arise long after exposure to distressing events. When there is impairment in social and/or occupational functioning, a clinical assessment is warranted. Combat and operational stress control is important to sustain Army strength over the long-term and reduce the cost to society, the DOD, the Soldiers, and families.

SECTION II — PRINCIPLES AND FUNCTIONAL AREAS FOR COMBAT AND OPERATIONAL STRESS CONTROL

COMBAT AND OPERATIONAL STRESS CONTROL INTERVENTIONS

1-17. Soldier and unit readiness is best achieved through an active, prevention-focused orientation that is embodied in the principles. These principles apply to all COSC interventions or activities throughout the theater and are followed by medical personnel in all BH/COSC elements. Their application may differ based on a particular level of care and other factors pertaining to the mission, enemy, terrain and weather, troops and support available, time available, and civil considerations (METT-TC).

1-18. Preventive interventions seek to reduce the occurrence or severity of COSR and behavioral disorders, thereby sustaining Soldier and unit readiness. These interventions are tailored to the needs of the population. There are four categories of preventive interventions that include—

- Universal. Interventions targeted to the general population or an assigned AO.
- Selective. Interventions targeted to a unit or Soldier whose risk is higher than average.
- Indicated. Interventions targeted to Soldiers with COSR or indications of a potential behavioral disorder, and to units that show signs their mission effectiveness is being affected by combat and operational stressors.
- Treatment. Interventions targeted to treat and follow-up Soldiers with behavioral disorders to prevent their loss from duty.

1-19. Identify life- or function-threatening medical, surgical, or psychiatric condition as soon as possible and provide those patients emergency treatment.

COMMUNICATIONS

1-20. To maximize prevention, COSC personnel must maintain a high degree of involvement with Soldiers and leaders of supported units. Regular visits to the battalion/squadron level are essential to maintain appropriate accessibility. The COSC personnel must always be attuned to Soldiers around them and not rely on Soldiers to come to them. This means that when time permits, COSC personnel need to start conversations with as many supported Soldiers as possible.

1-21. Communication is the delivery system of COSC services and therefore is essential. Use all available means to communicate/coordinate with supported units, subordinate teams, and higher headquarters to ensure delivery of effective and timely COSC and BH services.

COMBAT AND OPERATIONAL STRESS CONTROL MANAGEMENT PRINCIPLES

1-22. The COSC management principles of brevity, immediacy, contact, expectancy, proximity, and simplicity (BICEPS). These principles apply to all COSC interventions or activities throughout the theater, and are followed by COSC personnel in all BH/COSC elements. These principles may be applied differently based on a particular level of care and other factors pertaining to METT-TC.

> ***Note.*** The BICEPS principles are discussed in-depth in FM 6-22.5. This is a multiservice publication with the US Marine Corps (USMC). Due to differences in organizational structure the USMC defines the "C" in BICEPS to mean *centrality*, while the US Army defines it to mean *contact*, which will be used for the remainder of this manual.

Brevity, Immediacy, Contact, Expectancy, Proximity, and Simplicity

1-23. Using BICEPS is extremely important in the management of Soldiers with COSR and/or behavioral disorders.

Brevity

1-24. Initial rest and replenishment at COSC facilities located close to the Soldier's unit should last no more than 1 to 3 days (USMC and Navy is 3 to 4 days). Those requiring further treatment are moved to the next level of care. Since many require no further treatment, military commanders expect their Soldiers to RTD rapidly.

Immediacy

1-25. It is essential that COSC measures be initiated as soon as possible when operations permit. Intervention is provided as soon as symptoms appear.

Contact

1-26. The Soldier must be encouraged to continue to think of himself as a warfighter, rather than a *patient* or a *sick person*. The chain of command remains directly involved in the Soldier's recovery and RTD. The COSC team coordinates with the unit's leaders to learn whether the overstressed individual was a good performer prior to the COSR. Whenever possible, representatives of the unit or messages from the unit tell the Soldier that he is needed and wanted back. The COSC team coordinates with the unit leaders, through unit medical personnel or chaplains, any special advice on how to assure quick reintegration when the Soldier returns to his unit.

Expectancy

1-27. The individual is explicitly told that he is reacting normally to extreme stress and is expected to recover and return to full duty in a few hours or days. A military leader is extremely effective in this area of treatment. Of all the things said to a Soldier suffering from COSR the words of his small-unit leader have the greatest impact due to the positive bonding process that occurs. A simple statement from the small-unit leader to the Soldier that he is reacting normally to COSR and is expected back soon have positive impact. Small-unit leaders should tell Soldiers that their comrades need and expect them to return. When they do return, the unit treats them as every other Soldier and expects them to perform well.

Proximity

1-28. Soldiers requiring observation or care beyond the unit level are evacuated to facilities in close proximity to, but separate from the medical or surgical patients at the battalion aid station (BAS) or medical company nearest the Soldiers' unit. It is best to send Soldiers who cannot continue their mission and require more extensive intervention to a facility other than a hospital, unless no other alternative is possible. Combat and operational stress reactions are often more effectively managed in areas close to the Soldier's parent unit. On the noncontiguous battlefield characterized by rapid, frequent maneuver and continuous operations, COSC personnel must be innovative and flexible in designing interventions which maximize and maintain the Soldier's connection to his parent unit. See FM 6-22.5 for additional definition of proximity.

Simplicity

1-29. Indicates the need to use brief and straightforward methods to restore physical well-being and self-confidence.

1-30. The actions used for COSR control (commonly referred to as the 5 R's) involve the following actions:

- **R**eassure of normality.
- **R**est (respite from combat or break from the work).

- **R**eplenish bodily needs (such as thermal comfort, water, food, hygiene, and sleep).
- **R**estore confidence with purposeful activities and contact with his unit.
- **R**eturn to duty and reunite Soldier with his unit.

Note. Historically the Army had used the terms proximity, immediacy, expectancy and simplicity (PIES) but began using BICEPS when it became the approved joint terminology.

NONPATIENT STATUS

1-31. To prevent Soldiers with COSR from adopting the *patient* role, these guidelines should be followed:

- Keep the Soldier in uniform and hold him responsible for maintaining Soldier standards.
- Keep the Soldier separate from seriously ill or injured patients.
- Avoid giving him medications unless essential to manage sleep.
- Do not evacuate or hospitalize the Soldier unless absolutely necessary.
- Do not diagnose the Soldier prematurely.
- Transport the Soldier via general-purpose vehicles, not ambulances.

COMBAT AND OPERATIONAL STRESS CONTROL FUNCTIONAL AREAS

1-32. Combat and operational stress control interventions and activities are organized into nine functional areas. These functional areas cover the full spectrum of BH care from preventive through clinical intervention. They are defined below and are discussed further in subsequent chapters of this manual.

UNIT NEEDS ASSESSMENT

1-33. Unit needs assessment is the systematic and frequent assessment of supported units to determine the priority and types of BH interventions required. See Chapter 4.

CONSULTATION AND EDUCATION

1-34. Consultation involves the liaison with and preventive advice to commanders, staff of supported units, and Soldiers. Education involves training in concepts and skills for increasing Soldier resilience to stress. See Chapter 5.

TRAUMATIC EVENTS MANAGEMENT

1-35. Traumatic events management blends other COSC functional areas to create a flexible set of interventions specifically focused on stress management for units and Soldiers following potentially traumatizing events (PTE). Like other functional areas, COSC providers must tailor TEM to the needs of the unit and the Soldier. See Chapter 6.

RECONSTITUTION SUPPORT

1-36. Reconstitution is extraordinary action that commanders plan and implement to restore units to a desired level of combat effectiveness commensurate with mission requirements and available resources. It transcends normal day-to-day force sustainment actions. However, it uses existing systems and units to do so. No resources exist solely to perform reconstitution. In COSC reconstitution support, COSC personnel are responsible for providing unit Soldier restoration and conducting the COSC functions. This support is provided to units following traumatic events and during reconstitution, redeployment, and transition among levels of operational tempo (OPTEMPO). See Chapter 7.

Combat and Operational Stress Control Triage

1-37. Combat and operational stress control triage is the process of sorting Soldiers with COSR and/or BH disorders based upon where they can best be managed. See Chapter 8.

Combat and Operational Stress Control Stabilization

1-38. This function provides initial management of Soldiers with severe COSR or behavioral disorders. Their safety is ensured and they are evaluated for RTD potential or prepared for further treatment or evacuation. See Chapter 9.

Soldier Restoration

1-39. Soldier restoration involves the one- to three-day management of Soldiers with COSR or behavioral disorders normally near a medical treatment facility (MTF) in close proximity to his unit. This approach uses the 5 R's discussed above. See Chapter 10.

Behavioral Health Treatment

1-40. Patients with identified behavioral disorders receive ongoing evaluation, treatment and follow-up to sustain them. This functional area implies a therapist-patient relationship, clinical documentation, and adherence to clinical standards of care. See Chapter 11.

Soldier Reconditioning

1-41. Reconditioning is an intensive program of work therapy, military activities, physical training and psychotherapy. Reconditioning programs are conducted up to seven days (or more) in the corps area. Additional reconditioning may be provided in the theater outside the combat zone (CZ). See Chapter 12.

Priority of Functional Areas

1-42. The medical company/detachment, CSC commanders, MH section leaders, senior medical commanders, and command surgeons must set priorities, coordinate their actions, and allocate resources to accomplish missions based upon the total situation/METT-TC. Needs assessments and available resources guide COSC interventions. At the unit level, the unit needs assessment helps determine which functional areas (such as consultation and education; and critical event and transition management) take priority to achieve success with the COSC mission. At the Soldier level, COSC triage helps determine which functional areas (such as consultation and education, stabilization, Soldier restoration, behavioral health treatment, and reconditioning) take priority to achieve COSC mission goals.

Chapter 2

Behavioral Health and Combat and Operational Stress Control Elements in the Theater

SECTION I — MENTAL HEALTH SECTIONS

ASSIGNMENTS

2-1. Mental health sections are located in medical companies assigned to brigades, divisions, corps, and theater-level medical units.

ARMY OF EXCELLENCE

2-2. In the Army of Excellence (AOE) divisions, the division MH section (DMHS) is organic to the main support medical company (MSMC). In the armored cavalry regiments (ACRs) and separate brigades, a MH section is assigned to the forward support medical company (FSMC) of the forward support battalion (FSB) or a medical troop. In the corps and echelons above corps (EAC), the MH sections are located in the headquarters and headquarters detachment (HHD), area support medical battalion (ASMB) and in the area support medical companies (ASMCs). In those AOE divisions that are undergoing transformation, the MH sections are assigned to each of the brigade support medical companies (BSMCs) of the brigade support battalion (BSB) and in the ASMCs from the corps ASMB.

MODULAR FORCE

2-3. In the modular BCTs, Stryker BCTs (SBCTs) and the heavy BCTs (HBCTs) MH sections are assigned to the BSMC of the BSB. At echelons above brigade (EAB) and at the division, corps, or theater levels, the MH sections are assigned to the multifunctional medical battalion (MMB).

Note. Under the modular design, the MMB will replace the ASMB, the medical logistics battalion, and the medical evacuation battalion.

ORGANIZATIONS

2-4. The AOE divisions have only one MH section that is assigned to the MSMC of the main support battalion (MSB), which is located in the division support area (DSA). See Table 2-1 for a listing of personnel assigned to the division MH section. These personnel are deployed to the supported brigade as two-man teams comprised of one officer and one MH specialist.

Table 2-1. Division mental health staff

Division Psychiatrist (MAJ, Area of Concentration [AOC] 60W00, Medical Corps [MC])
Social Work Officer (CPT, AOC 73A67, Medical Service [MS] Corps)
Clinical Psychologist (CPT, AOC 73B67, MS)
Mental Health NCO (E7, E6, and E5, Military Occupational Specialty [MOS] 68X40/30/20)
Mental Health Specialist (Four)—(Two E4 and Two E3, MOS 68X10)

FUNCTIONS AND RESPONSIBILITIES OF ALL MENTAL HEALTH SECTIONS

2-5. All MH sections regardless of their organizational assignment are tasked with providing COSC for their supported units. In all of these units, COSC is accomplished through vigorous prevention, consultation, training, educational, and Soldier restoration programs. These programs are designed to provide BH expertise to unit leaders and Soldiers where they serve and sustain their mission focus and effectiveness under heavy and prolonged stress. The MH sections identify Soldiers with COSRs who need to be provided rest/Soldier restoration within or near their unit area for rapid RTD. These programs are designed to maximize the RTD rate of Soldiers who are either temporary impaired, have a diagnosed behavioral disorder, or have stress-related conditions. Also the prevention of posttraumatic stress disorders (PTSDs) is an important objective for brigades and EAB. The behavioral health officer (either a clinical psychologist or social work officer) and MH specialist are especially concerned with assisting and training of—

- Small-unit leaders.
- Unit ministry teams (UMTs) and staff chaplains.
- Battalion medical platoons.
- Patient-holding squad and treatment squad personnel of the medical company.

RESPONSIBILITIES

2-6. The MH section has a primary responsibility for assisting commanders with COSC by implementing the brigade combat mental fitness program. Also the MH section serves as a consultant to the commander, staff, and others involved with providing prevention and intervention services to unit Soldiers and their families. The MH section has staff responsibilities for assisting the brigade surgeon with establishing brigade policy and guidance for the prevention, diagnosis, treatment, management, and RTD of stress-related casualties. This is accomplished under the guidance and in close coordination with maneuver battalions and FSMC/BSMC physicians.

FUNCTIONS

2-7. The COSC functional areas were identified in Chapter 1. The functions of all MH sections are to support the nine functional areas except Soldier reconditioning.

UTILIZATION IN GARRISON

2-8. In garrison, BH personnel assigned to the division, brigade, or corps/EAC units continue to perform the same staff and outreach functions with supported units as they do in a field environment. An increase in the BH treatment functions may be possible as a result of consolidating BH care providers. The BH providers make available their consultation skills and clinical expertise to the Soldiers of supported units and their family readiness groups. Clinical care of family members and of Soldiers that require longer-term care beyond crisis intervention, brief treatment, and medication follow-up is the responsibility of the medical department activity (MEDDAC)/medical center (MEDCEN). The MH section personnel should focus their clinical work primarily on Soldiers with problems amenable to brief treatment.

2-9. Clinical services may be provided as part of a consolidated BH activity (as is usually established by a DMHS) or by augmenting an existing MEDDAC/MEDCEN BH staff.

2-10. The MH sections should strive to reduce referrals for BH treatment by working closely with unit leaders and chaplains to control organizational stress and rapidly identify and intervene with those Soldiers having BH disorders.

2-11. When the medical company or battalion deploys on training exercises, assigned BH personnel deploy with them to provide COSC training and support. In addition, they train to improve their own technical and tactical skills.

Note. In accordance with AR 40-216, clinical responsibilities in garrison must not interfere with participation in field and deployment exercises and maintenance of combat readiness.

MEDICAL DETACHMENT, COMBAT STRESS CONTROL

MISSION

2-12. The mission of the medical detachment, CSC (TOE 08463A000) is to provide COSC interventions and stress prevention activities to supported units in its AO. The staff augments division and brigade MH sections; provides direct support (DS) to combat brigades without organic BH officers; and provides area support in its AO. The medical detachment, CSC reconstitutes other brigade and division COSC assets. The medical detachment, CSC provides COSC interventions and activities to indigenous populations as directed in stability and reconstruction operations, humanitarian assistance, disaster relief, peace support operations, and detention facility operations. The medical detachment, CSC provides COSC interventions and activities to units in support of their readiness preparation and throughout their deployment cycle.

ASSIGNMENT

2-13. The Medical Reengineering Initiative (MRI) medical detachment, CSC (TOE 08463A000) is assigned to a corps medical brigade (MEDBDE) (TOE 08422A100), or other medical command and control (C2) element. Its teams may be further attached to a medical company, CSC (TOE 08467L000), an ASMB (TOE 08456A000), a BSMC/MSMC, or a combat support hospital (CSH). See Appendix B for the proposed future medical detachment, COSC.

EMPLOYMENT AND CAPABILITIES

2-14. The detachment headquarters is usually located in the DSA. Its teams disperse throughout the division's AO, and may extend support to corps if no CSC medical company is available. At TOE Level 1, the MRI medical detachment, CSC provides—

- A preventive section capable of dividing into four preventive teams. Each team provides mobile COSC interventions and activities.
- A fitness section capable of dividing into two fitness teams. Each fitness team provides mobile COSC interventions and activities. Each team is equipped to hold 40 Soldiers at the same time. With additional logistical support, each team can accommodate additional Soldiers under surge conditions.

STAFF RESPONSIBILITIES

2-15. The medical detachment, CSC assists the C2 headquarters (of the unit to which it is assigned or attached) regarding planning and coordination of COSC support, stress threat, mental and physical stressors, stress behaviors, principles of COSC, and implementation of COSC functional areas.

DEPENDENCY

2-16. This unit is dependent on—

- Appropriate elements of the corps for FHP, religious support, legal, finance, field feeding, personnel and administrative services support; laundry and clothing exchange; mortuary affairs support; and security of enemy prisoners of war (EPW), detainee, and US prisoner patients.
- The medical headquarters to which it is assigned/attached for FHP; medical administration, logistics, including MEDLOG; medical regulating of patients; evacuation; coordination for RTD; and unit level equipment and communications-electronics (CE) maintenance.

MOBILITY AND SECURITY

2-17. This unit is 100-percent mobile. It requires 100 percent of its organic personnel and equipment be transported in a single lift, using its organic vehicles. Upon relocation, Soldiers being held will require additional transportation. This unit is responsible for perimeter defense of its immediate operational area. However, it is dependent on appropriate elements of the corps for additional security, to include security of convoy operations. Personnel of the detachment are provided weapons for their personal defense and for the defense of their patients and/or held Soldiers.

ORGANIZATION

2-18. This 43-person MRI medical detachment (Figure 2-1) is organized into a headquarters section; a preventive section composed of four CSC preventive (CSCP) teams; and a fitness section consisting of two CSC fitness (CSCF) teams.

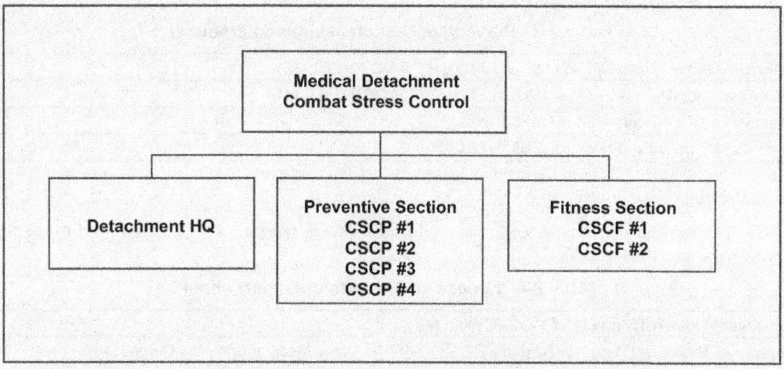

Figure 2-1. Medical Reengineering Initiative medical detachment, combat stress control (TOE 08463A000)

DETACHMENT HEADQUARTERS

2-19. The medical detachment, CSC headquarters provides C2 for the detachment. The headquarters element is responsible for planning, coordinating, and implementing COSC support for supported units. Personnel of the headquarters element provide maintenance, supply and service, and personnel administrative support. See Table 2-2 for a list of personnel assigned to the detachment headquarters.

Table 2-2. Detachment headquarters assigned personnel

Detachment Commander (LTC, AOC 60W00, MC)
Field Medical Assistant (1LT, AOC 70B67, MS)
Detachment Sergeant (E7, MOS 68X40)
Supply Sergeant (E5, MOS 92Y20)
Human Resource Specialist (E4, MOS 42A10)
Wheeled Vehicle Mechanic (E4, MOS 63B10)
Cook (E3, MOS 92G10)
Note. Detachment officers and NCOs from the preventive section and the fitness section may be assigned additional duties, which enhance the overall effectiveness of the headquarters section.

Preventive Section

2-20. The medical detachment, CSC has a 16-person preventive section. See Table 2-3 for listing of personnel assigned to the preventive section. The section can be divided into four separate CSCP teams. Three CSCP teams are normally allocated to supported maneuver brigades (one team per brigade). The fourth CSCP team may provide direct support to a SBCT, an aviation brigade, other brigade-sized units, or to corps units operating in a division's AO. Preventive section personnel may be task-organized with personnel of the fitness section into teams for specific missions. The section (and team) leader position may be held by any of the officers assigned to the section. The section's COSC interventions and activities are—

- Unit needs assessment; consultation and education; TEM; COSC triage; stabilization (emergency); and BH treatment.
- Assisting with Soldier restoration and reconditioning at the CSC detachment program. Overseeing a one- to three-day COSC Soldier restoration program in a brigade, division, or ASMC holding section or in another area suitable for Soldiers experiencing COSR and/or other stress-related disorders.

Table 2-3. Preventive section assigned personnel

Social Work Officer (Four)—(Two MAJ and Two CPT, AOC 73A67, MS)
Clinical Psychologist (Four)—(Two MAJ and Two CPT, AOC 73B67, MS)
Mental Health NCO (Four)—(E5, MOS 68X20)
Mental Health Specialist (Four)—(E4, MOS 68X10)

Fitness Section

2-21. The medical detachment, CSC has a 20-person fitness section. Personnel assigned to the fitness section are identified in Table 2-4.

Table 2-4. Fitness section assigned personnel

Psychiatrist (Two)—(LTC and MAJ, AOC 60W00, MC)
Occupational Therapist (Two)—(MAJ and CPT, AOC 65A00, Army Medical Specialty Corps [SP])
Psychiatric/BH Nurse (Two)—(MAJ and CPT, AOC 66C00, Army Nurse Corps [AN])
Occupational Therapy NCO (Four)—(Two E6 and Two E5, MOS 68W30/20N3)
Mental Health NCO (Four)—(Two E6 and Two E5, MOS 68X30/20)
Mental Health Specialist (Six)—(Four E4 and Two E3, MOS 68X10)

2-22. The fitness section is task-organized to provide COSC interventions and activities. Fitness section personnel may be task-organized with personnel of the preventive section into teams for specific missions. This section can divide into two CSCF teams. Each fitness team can deploy a four-person mobile team using the team's high-mobility multipurpose wheeled vehicle (HMMWV). One CSCF team usually collocates with a supported divisional medical company to provide mobile COSC support within a DSA and conduct Soldier restoration programs, as required. The second CSCF team augments area support to corps units in the division AO and in forward areas of the corps. The CSCF team provides staff and equipment for operating a Soldier restoration or reconditioning center. This section's COSC interventions and activities are—

- Unit needs assessment; consultation and education; TEM; COSC triage; stabilization; and BH treatment.
- Conducting Soldier restoration and reconditioning programs.
- Staffing and operating a psychiatric ward for supported CSH when this capability is required (refer to FM 8-10-14).

MEDICAL COMPANY, COMBAT STRESS CONTROL

2-23. The MRI medical company, CSC is employed in the corps and EAC. The basis of allocation is one medical company, CSC per corps or theater. The medical company, CSC is task-organized, METT-TC-dependent for stability and reconstructions operations. Medical company, CSC (TOE 08467A000) replaces medical company, CSC (TOE 08467L000). A medical company, CSC provides comprehensive preventive and treatment services to a corps and EAC during war. It provides this support to all services on an area support basis. The medical company, CSC provides DS to combat and combat support (CS) brigades, as needed. It reinforces or reconstitutes other CSC assets in the corps or divisions if required. The medical company, CSC provides COSC/BH services to indigenous populations as directed in stability and reconstruction operations, humanitarian assistance, disaster relief, and peace support operations. The comprehensive support provided by the CSC medical company entails all of the nine COSC functional mission areas. The COSC functional mission areas are identified in Chapter 1 and discussed in subsequent chapters of this publication.

CAPABILITIES

2-24. At TOE Level 1, the CSC medical company provides—

- Advice, planning, and coordination for COSC to commanders.
- Reconstitution (COSC) support for units up to division size.
- Preventive and fitness teams (four to ten personnel) for consultation, treatment services, and reconstitution support for up to battalion-sized organizations.
- Soldier restoration or reconditioning programs for up to 50 Soldiers per fitness team on an area basis.
- Deployment of COSC elements to forward areas for support of contingency operations.

ASSIGNMENT

2-25. The medical company, CSC (TOE 08467A000) is assigned to a Medical Command (US Army) (MEDCOM). Elements of this TOE may be further attached to a corps MEDBDE or to either a MMB or an ASMB.

ORGANIZATION

2-26. The medical company, CSC is organized into a headquarters section, a preventive section, and a fitness section. The company is dependent on appropriate elements of the MEDCOM or MEDBDE for administrative and MEDLOG support, medical regulating, COSR casualty delivery, and medical evacuation. The company is dependent on appropriate elements of the corps or EAC for finance, legal, personnel and administrative services, food service, supply and field services, supplemental transportation, and local security support services. When CSC medical company elements or teams are deployed to division areas, they are dependent on the division medical companies (such as MSMC or FSMC/BSMC) for patient accounting, transportation, food service, and field service support.

Employment

2-27. The medical company, CSC is employed in all intensities of conflict when a corps with two or more divisions is deployed. Task-organized COSC elements are deployed for division-sized combat operations, stability and reconstruction operations, and other contingency operations which are METT-TC-dependent. Together, the preventive and fitness sections provide all five BH disciplines. These resources are flexibly task-organized in a variety of combinations to meet the fluid COSC threat at different phases in the operations. Personnel may be quickly cross-attached from one section to another to accommodate the shifting workload and to provide reconstitution support packages. The preventive and fitness sections both organize into teams. The COSC preventive or fitness teams deployed forward of the corps boundaries in support of tactical operations come under the control of the FHP operations element in the supported units. These teams will also come under the operational control of the division or brigade CSC teams. One or more of the medical company, CSC's eight preventive teams may locate at the BSMC when deployed in DS of SBCT, HBCT, and separate brigades or ACRs.

Fitness Teams

2-28. One or more of the four (ten-person) fitness teams may reinforce ASMCs which are deployed to locations throughout the corps and EAC. These teams provide a basis for COSC prevention and intervention. The teams may conduct Soldier restoration programs at the ASMCs, as required. These teams may also be deployed forward to provide temporary augmentation/reinforcement, as required.

2-29. Based on workload, one or more of the four fitness teams, plus one or more preventive teams locate with a theater hospital where they conduct Level IV COSC reconditioning programs, as required. A hospital located in the corps rear or out of the CZ is the best location to conduct the theater COSC reconditioning program. When deployed with a hospital, these teams provide mobile consultation in the vicinity of the hospital. These teams are also prepared to restrict reconditioning programs and deploy forward in support of higher priority missions on very short notice. These teams can also augment hospital BH ward services by staffing a temporary BH ward.

2-30. The medical company, CSC is divisible into four functionally emulative increments for split-based operations and stability and reconstruction operations as assigned. Nonstandard task elements for specific missions can be organized using any combination of the preventive section and fitness section personnel to meet specific BH needs. For stability and reconstruction operations, the minimum is an officer/NCO team to supplement a brigade COSC team or a preventive module/team of two officers, one NCO, and one enlisted. These modules may be augmented with personnel from the fitness section to add additional specialty expertise.

Headquarters Section

2-31. The headquarters section provides C2 and unit-level administrative and maintenance support to its subordinate sections when they are collocated with the company. The headquarters section may also provide assistance to detached elements by making site visits if the elements are within a feasible distance for ground transportation. The CSC medical company elements normally deploy with limited maintenance capability. When these COSC elements deploy, they are dependent on the supported units for patient accounting, transportation, food service, and field services. The personnel assigned to the headquarters section are identified in Table 2-5. See Appendix C for information on the AOE L-Series TOE medical company and medical detachment, CSC.

Table 2-5. Medical company, combat stress control, headquarters section personnel

Commander (LTC, AOC 60W00, MC)
Chaplain (CPT, AOC 56A00, Chaplain [CH])
Field Medical Assistant (CPT, AOC 70B67, MS)
First Sergeant (E8, MOS 68W50)
Mental Health NCO (E7, MOS 68X40)
Chemical Operations Specialist (E5, MOS 54B20)
Patient Administrative NCO (E5, MOS 68G20)
Supply Sergeant (E5, MOS 92Y20)
Power Generator Equipment Repairer (E4, MOS 52D10)
Wheeled Vehicle Mechanic (E4, MOS 63B10)
Administrative Specialist (E4, MOS 71L10)
Patient Administrative Specialist (Two)—(E4 and E3, MOS 68G10)
Armorer (E4, MOS 92Y10)
Cook (E3, MOS 92G10)
Note. Personnel from the headquarters section are deployed with teams or task-organized COSC element, as required.

SECTION II — OPERATIONAL STRESS ASSESSMENT TEAM IN A THEATER

MISSION OF OPERATIONAL STRESS ASSESSMENT TEAMS

2-32. The operational stress assessment team (OSAT) is task-organized when needed from personnel in the research institutes of United States Army Medical Research and Materiel Command (USAMRMC). The mission of this team is to deploy with BH research expertise, equipment and supplies to provide scientifically valid needs assessment of multiple units in the theater and macro analysis of COSC stressors, stress, and Soldier needs. The source is predominantly the Walter Reed Army Institute of Research, augmented when needed with research psychologists and enlisted technicians from the US Army Aeromedical Research Laboratory, the US Army Medical Research Institute of Chemical Defense, and the US Army Research Institute of Environmental Medicine. The OSATs are USAMRMC teams that are assembled entirely from personnel in table of distribution and allowances (TDA) units and are dependent on TOE units when they deploy to a theater.

ASSIGNMENT AND CAPABILITIES

2-33. The OSAT normally is used for augmenting the senior medical command in the theater. The OSAT has the capabilities to—

- Plan, coordinate, and conduct questionnaire and interview surveys.
- Establish a database and analyze scientific-quality data from their own surveys and from standardized unit assessments conducted by unit MH section and CSC units for their local supported units.
- Provide rapid reports of findings to the command that include recommendations for policy and for COSC actions and interventions.
- Design and validate survey instruments, including ones to address new and emerging stressors and issues in the theater.
- Use USAMRMC capability to archive and analyze data for comparison with other conflicts.
- Educate and mentor COSC and other personnel in survey methodology and provide them, as well as the theater commander, the wider perspective to use in their local UNA.

- Interface with the theater surveillance capabilities of preventive medicine (PVNTMED) and occupational and environmental medicine to address the BH aspects of chemical, biological, radiological and nuclear (CBRN) exposures and threats.

DEPENDENCY

2-34. The OSAT depends on the medical headquarters or the units with which it is working for administrative support. The OSAT depends on the theater (line) command for implementation of the approved surveys.

MOBILITY AND SECURITY

2-35. The OSAT depends on the medical and theater command and the units with which they work for transportation. The OSAT personnel are issued personal weapons and protective equipment, but depend on the units they are with for protection.

COMMUNICATIONS

2-36. The OSAT depends on the medical and theater command and the units with which they work for communications. With approval of theater command, they may be allowed USAMRMC-issued phones for satellite communication for mission-essential purposes.

ORGANIZATION

2-37. The typical OSAT has three officers and two enlisted Soldiers. The officer positions are listed as research psychologists (AOC 71H), but may be filled by clinically qualified BH officers with research training and experience. The enlisted are usually MH specialists (MOS 68X) with research training and experience. Staffing may also include personnel with information management specialties. The equipment is the required personal equipment of the team members, plus data collection and automated data processing equipment appropriate to the mission.

Chapter 3

Combat and Operational Stress Control Support Operations

SECTION I — COMBAT AND OPERATIONAL STRESS CONTROL PROFESSIONAL DISCIPLINES AND PROFESSIONAL CONSULTANTS

PROFESSIONAL DISCIPLINES

3-1. There are five BH professional disciplines and two enlisted specialties that serve in support of the COSC mission. The professional disciplines and enlisted specialties were identified in Chapter 2. The professional disciplines include social work, clinical psychology, psychiatry, occupational therapy (OT), and psychiatric/BH nursing; the enlisted specialties are in BH and OT. While much of the COSC knowledge base and most of the skills are shared by all the BH personnel, each discipline brings its own perspective from its professional training, skills that can only partially be familiarized to the others, and in some cases unique credentials to conduct specific assessments and treatments. The special perspectives and skill sets of the officer and enlisted members of the COSC team are summarized in the following subparagraphs.

OFFICER AREAS OF CONCENTRATION

Social Work Officer

3-2. The social work officer (AOC 73A67) evaluates the psychosocial systems affecting the Soldier, his family, and the unit; identifies and coordinates with military and civilian support agencies (including the Army Substance Abuse Program and the Family Advocacy Program [FAP]); assesses BH disorders; provides a range of counseling and psychotherapies. Installation MTFs are responsible for BH services that are beyond the scope of COSC functional areas such as long-term clinical care, inpatient services, and alcohol treatment programs. This officer provides proactive consultation, conducts individual and group counseling, supervises Soldier restoration/reconditioning, and coordinates RTD of recovered cases. He also provides staff advice and coordinates Army and civilian social services support.

3-3. When deployed as a member of a COSC preventive team or task-organized COSC element, the social work officer's duties include—

- Evaluating psychosocial (unit and family) functioning of Soldiers with COSR and misconduct stress behavior.
- Coordinating and ensuring the return of recovered COSR and neuropsychiatric (NP) Soldiers to duty and their reintegration into their original or new unit.
- Identifying and resolving organizational and social environmental factors which interfere with combat readiness.
- Coordinating support for Soldiers and their families through Army and civilian community support agencies, when possible.
- Apprising unit leaders, primary care physicians, and others health care providers of available social service resources.
- Providing consultation to supported unit commanders and to other BH/COSC personnel regarding problem cases.
- Counseling and providing therapy for Soldiers with psychological problems.
- Conducting and supervising unit needs assessment and TEM.

3-4. The doctoral level social work officer, as outlined in DODD 6490.1, may conduct clinical evaluations for imminent risk or dangerousness.

Clinical Psychologist

3-5. A clinical psychologist (AOC 73B67) diagnoses BH disorders and a range of psychotherapies including cognitive/mental approaches, assessment for safety evaluations, and psychometric assessment.

Psychiatrist

3-6. The psychiatrist (AOC 60W00) is a physician that is credentialed to diagnose and treat BDPs. He also conducts mental assessments, makes differential diagnosis, provides a range of psychotherapies and safety evaluations, and prescribes medication.

Psychiatric/Behavorial Health Nurse

3-7. This AN officer (AOC 66C007T) provides specialized nursing services for emotionally distressed individuals and promotes COSC within the MTF and the adjacent military community. He performs liaison and consultative functions to ensure continuity of patient care. The additional skill identifier of 7T indicates that advanced skills and competencies are required for this position and that the individual may act as a consultant in clinical nursing practices.

Occupational Therapist

3-8. The occupational therapist (AOC 65A00), plans, implements, and supervises preventive and restorative OT services to enhance the occupational performance of Soldiers. This officer evaluates Soldier's performance across the spectrum of occupational areas (such as activities of daily living, work, education, leisure, and social participation). Analyze jobs and job tasks for underlying performance requisites. Identify and evaluate mental and physical stressors and stress reactions and teach prevention, adaptive coping and psychosocial skills. Apply therapeutic media in both, individual and group settings to include environmental adaptation, the therapeutic use of self, consultation, education, and occupation (purposeful activity). Match Soldiers to therapeutic and/or vocational activities based on therapeutic need and functional ability.

ENLISTED MEDICAL OCCUPATIONAL SPECIALTIES

MENTAL HEALTH SPECIALIST

3-9. The MH specialist (MOS 68X30/20) provides the perspective of the enlisted Soldier, takes the initial history and conducts mental status evaluations for supervising officer, and administers psychological tests. The MH specialist also performs counseling and may have advanced specialty training (such as drug and alcohol counseling with higher ranked individuals).

OCCUPATIONAL THERAPY SPECIALIST

3-10. The OT specialist (MOS 68W30/20N3) provides the perspective of the enlisted Soldier, assists the supervising OT officer in evaluating a Soldier's occupational performance; conducts initial occupational performance history interviews and mental status evaluations; observes clients to gather data as part of task performance skill assessments; and implements OT interventions under the supervision of an OT officer.

CONSULTANTS

3-11. There are two levels of COSC consultants—the theater-level consultant and the subordinate commander consultant. The theater-level COSC consultant is normally the senior COSC officer designated by the theater surgeon and approved by the theater commander. The subordinate command COSC consultants are MH officers that function as the point of contact (POC) for one or more commanders.

These consultants oversee, coordinate, and conduct COSC interventions and activities for their respective commanders. They accomplish the following tasks appropriate to their level of operation:

- Direct and report AO-wide COSC needs assessments.
- Coordinate AO-wide implementation of COSC functional areas.
- Ensure COSC standards are established and consistently applied.
- Ensure compatibility between COSC standards and hospital BH standards (theater/AO COSC consultant).
- Define COSC operational requirements.
- Recommend the AO COSC evacuation policy (theater/AO COSC consultant).
- Prepare the COSC portion of the FHP estimate and operations plan.
- Review all command medical policies affecting COSC interventions and activities.
- Evaluate the quality of COSC interventions and activities rendered in the AO.
- Make regular AO COSC consultant site visits to COSC sections and units.
- Coordinate with other COSC consultants in adjoining AOs and in the primary medical support elements to ensure unity of effort.
- Oversee AO COSC training.
- Coordinate joint and combined COSC interventions and activities as necessary.
- Plan for future COSC operations.
- Consult with the command surgeon on all COSC matters.

SECTION II — DEPLOYMENT AND EMPLOYMENT OF MENTAL HEALTH SECTIONS

MENTAL HEALTH SECTIONS

3-12. The MH section assignments, staff, functions and responsibilities were discussed in Chapter 2. The information provided in this section pertains to deployment and employment and how these sections prepare for COSC support for their units in garrison, combat operations, and stability and reconstruction operations predeployment. The involvement of MH section with support units during field and other training exercises is important. It permits BH personnel to gain a familiarity with type of units supported, how they conduct operations, their mission, and the likely stressors associated with a particular type of unit. It also allows them to interact and become familiar with unit personnel and to gain some level of trust through their interactions. Combat and operational stress control personnel should be active participants in all training exercises. Unit predeployment training must be planned when the unit is alerted for its deployment but scheduled after cross-leveling and filling of all positions is completed. Training of BH personnel will include—

- Ensuring their active participation in the normal training cycle of the BCT/ACR to which assigned.
- Training the BH officers and specialists in all COSC functional areas that support their unit.
- Working with the supported units on a frequent basis so BH personnel may gain a familiarity with the unit's personnel, capabilities, and possible stressors associated with its mission.
- Providing predeployment train-up to include refresher and operation-specific training before deployment.

PLANNING

3-13. Since planning is normally performed at battalion level, MH sections are involved in developing a COSC estimate to support the unit plan. The COSC estimate of the situation is derived from mission analysis, to include considerations for COSR, BH disorders, substance abuse, misconduct behaviors, suicide, major stressors, troop populations to be supported, and COSC augmentation requirements or assets available.

Coordinating

3-14. The BH personnel may be involved in coordinating COSC support with supporting and supported units, as appropriate. Brigade BH personnel must be known and trusted by the leaders of the units they support. Brigade BH personnel coordinate with other BH officer and enlisted personnel pertaining to the professional disciplines associated with conducting the COSC mission. They also coordinate with other key medical and nonmedical personnel (such as health care providers, surgeons, chaplains, UMTs that assist or may be involved with the COSC mission).

Deployment of Mental Health Sections

Early Deployment

3-15. The MH section should be deployed early into staging areas with the lead elements of the medical company and supported units.

Command and Control

3-16. The MH section falls under the C2 of the FSMC/BSMC and may receive mission guidance from the brigade surgeon through the FSMC/BSMC commander. Tasking for COSC operational support flows through the BSMC/FSMC headquarters. When BH personnel from the medical detachment/company, CSC are attached, the MH section ensures the timely flow of operations and situation awareness information to that element.

Employment of the Mental Health Sections

3-17. Employment of the MH section is based on the type of operations being conducted. The MH section normally operates from the FSMC/BSMC located in the brigade support area (BSA). When MH personnel are deployed forward, they are dependent on supported units for rations and logistical support.

Mental Health Functional Area Considerations

3-18. When determining COSC support for BCT units, the BH team should review each of the BH functional areas and determine which BH functions need to be implemented for each supported unit. Also, in developing the COSC plan, the BH team should consider—

- Conducting a UNA for each supported battalion and separate company supported.
- Establishing a consultation and education/training program for supported units to include—
  - Traumatic event and transition management.
  - The need for COSC triage, stabilization, and BH treatment should be addressed as they arise.
  - Soldier restoration, which can only be accomplished within and by the BSMC holding section or with augmentation by or referral to CSC unit personnel.
  - Reconditioning services that will need to be referred to nearest location where available.

SECTION III — SUPPORT OPERATIONS CONDUCTED BY THE MEDICAL UNIT, COMBAT STRESS CONTROL

PREDEPLOYMENT ACTIVITIES FOR ALL MEDICAL UNITS, COMBAT STRESS CONTROL

3-19. The predeployment activities identified in Section II for MH sections also apply for the units in Section III that include the ASMB or MMB and medical companies and detachments, CSC. These units must also keep themselves focused on applying the nine functional areas during the predeployment phase.

Note. The MMB is the C2 organization which will replace the ASMB, the medical logistics battalion, and the evacuation battalion.

MEDICAL DETACHMENT, COMBAT STRESS CONTROL DEPLOYMENT

3-20. The medical detachment, CSC is deployed to provide COSC support whenever a division or two BCTs/regiments are deployed. It is assigned to a medical brigade or other medical headquarters for C2. Elements of the detachment may be further attached to FSMCs/BSMCs for operational control and logistical support while in support of the BCT.

MEDICAL COMPANY, COMBAT STRESS CONTROL DEPLOYMENT

3-21. The primary mission of this unit is to provide COSC support throughout a corps AO. It is assigned to either a MEDCOM/MEDBDE and may support from two or up to five divisions. They assign teams to any operational area based on troop populations without BH support. They move teams rapidly to areas of surge workload or mass casualties.

MEDICAL FORCE 2000 VERSES THE MEDICAL REENGINEERING INITIATIVE DETACHMENTS

3-22. The Medical Force 2000 (MF2K) and MRI CSC detachments conduct COSC support operations very similarly. However, in the MRI detachment, psychiatry assets are assigned to the fitness section and psychology assets are assigned forward to the prevention section as compared to the MF2K detachment, the manning strategy for psychology and psychiatry is reversed. The psychologists are assigned to the fitness section and psychiatrists are assigned to the prevention teams.

Medical Detachment, Combat Stress Control Support to a Division and Brigade Combat Teams During Combat Operations

3-23. When in support of an AOE division, a medical detachment, CSC is usually attached to the MSMC/BSMC and for a transformation division the medical detachment, CSC is attached to the MMB or MEDBDE. It can be attached to other battalion-sized units for support. It is under the operational control of the support battalion and MSMC/FSMC/BSMC and closely coordinates its augmentation support with the division surgeon and division psychiatrist. Habitual long-term relationships of medical detachments, CSC with specific divisions should be established so that COSC support is integrated as a normal function of the FHP mission. However, as an echelon above division (EAD) asset, the detachment (or its modular teams) may be task-organized and deployed to support other units or missions other than divisional units, as workloads require. The medical detachment, CSC depends on the units to which it is attached for administrative and logistics support. It may be reinforced and/or receive personnel replacements from a medical company, CSC or from another medical detachment, CSC.

Preventive Section Operations

3-24. Upon arrival to a theater, the detachment is initially attached to a higher medical headquarters for C2 and its three CSCP teams of four-person each are usually further attached. One CSCP is attached to each divisional MSMC prior to commencement of combat operations. This will allow the teams to linkup with and augment division COSC assets in support of the brigades.

- The deployment of each CSCP provides additional COSC personnel that increase COSC augmentation capabilities at the FSMC. A CSCP may routinely visit the BASs to provide consultation. The CSCP may deploy, as needed, to provide reconstitution support to units undergoing hasty or deliberate reorganization.
- In some operations, some or all of the CSCP personnel may either be attached to an ASMC or are deployed in support of a maneuver brigade or reconstitution site. It is important for all CSCP personnel to understand that they are responsible for all functional areas, and will conduct them as dictated by the mission.

Fitness Section Operations

3-25. The CSCF may locate with any Level II or above MTF in the AO. This section has tents and is equipped to operate a Soldier restoration center and can supervise Soldier restoration activities to include consultation, rest, relaxation, and treatment as necessary for speedy RTD within three days. The CSCF also provides triage and stabilization at the MSMC and consultation to other division units located in the DSA and division rear areas. All CSCF members are responsible for providing support for all COSC functional areas as dictated by the mission.

3-26. The CSCF staff may move within the AO to temporarily reinforce or reconstitute a CSCP at the brigade level, or to escort COSR casualties to the Soldier restoration center. The CSCF personnel give reconstitution support to attrited units, especially when the units return to the DSA. The CSCF supplements the division psychiatrist; brigade COSC teams, CSCPs, and chaplains and leaders in after-action debriefings. The CSCF helps integrate recovered Soldiers and new replacements into units during reorganization activities. When either the BSA or DSA is tactically too unstable to allow Soldier restoration operations, the CSCF may locate with an ASMC or a CSH and attempt to continue their support to the division from that location.

COMBAT STRESS CONTROL DETACHMENT IN CONTINGENCY OPERATIONS

3-27. For contingency operations, a task-organized CSC element from the CSC medical detachment may be deployed. A CSCP supporting a maneuver brigade may operate out of a central base of operations. The team may deploy forward to the BSA (base camps or fire bases), in response to—

- Anticipated battle.
- After-action debriefing requirements.
- Alcohol/drug problems in a unit.
- Incidents of misconduct stress behaviors.
- High incidence of COSRs.
- Unit rotation in or out of theater.

3-28. Several CSCF and CSCP from two or more CSC medical detachments may be consolidated under the command of the senior medical detachment, CSC commander to staff a central reconditioning program for the AO. This may also function as an alcohol/drug detoxification rehabilitation program. This may be referred to as a COSC center and will also provide consultation and treatment support to MP confinement facilities where misconduct stress behaviors may have led to incarceration. This could also include support for EPW and detainee confinement facilities.

STABILITY AND RECONSTRUCTION OPERATIONS

3-29. In stability and reconstruction operations, if the force deployed is smaller than a division, a medical detachment, CSC would not be required to provide COSC support. In such cases, the medical company, CSC or the medical detachment, CSC may be tasked with providing either a COSC team or a task-organized COSC element. The COSC team or the task-organized COSC element is attached to the supporting medical headquarters or to an MTF and conducts its mobile consultation mission.

MEDICAL UNITS, COMBAT STRESS CONTROL INTERFACE AND COORDINATION REQUIREMENTS

3-30. The medical units CSC must interface with its higher headquarters element and with the unit to which it is attached. The higher headquarters may be a MEDCOM, MEDBDE, or an ASMB/MMB.

- Interface between the CSC units and their higher headquarters. The medical unit CSC interfaces with its higher headquarters pertaining to its assigned mission. It provides estimates and has input to the operation plan (OPLAN). The unit receives its OPORD from the higher headquarters. Interface between the unit and higher headquarters staff elements will include the following subject areas—

- The COSC operations.
- Assignment or attachment of the medical detachment, CSC elements.
- Daily personnel and equipment status reports.
- Class VIII (medical supply) status and resupply requirements.
- Casualty feeder reports.
- Operation orders.
- Personnel replacement for the detachment.
- Medical intelligence information.
- Behavioral health/COSC consultation tasking and results.
- Maintenance requirements and requests.
- Replacement and reconstitution operations.
- Civil-military operations.
- Communications (signal operation instructions [SOI], access to message centers and nets, and transmission of COSC messages through medical, land, and other channels).
- Mass casualty plan.
- Road movement clearances.
- Tactical updates.
- Contingency operations.
- Return-to-duty and non-RTD procedures.
- Medical evacuation procedures (air and ground ambulances).
- Changes in locations of supported unit.

- Interface and coordination with unit to which attached. The headquarters of the unit to which a medical detachment, CSC is under operational control (OPCON) or attached is responsible for providing the administrative and logistical support requirements of the detachment. These requirements are normally identified in the attachment order. The higher headquarters will be identified in the attachment order and the medical detachment, CSC will be coordinated with this headquarters prior to deployment. The medical detachment, CSC must coordinate with the headquarters staff according to the tactical standing operating procedures (TSOP) of the unit of attachment. The staff shares information with the detachment commander or his representative pertaining to the threat, tactical situation, patient/COSR casualty status, and changes in FHP requirements. Coordination activities and subject area information exchange should include—
  - Command and control procedures.
  - Status of FSMCs/BSMCs and CSCPs.
  - Communications and SOI.
  - Operational support requirements.
  - Civil-military operations.
  - Soldier restoration operations.
  - Reinforcement and personnel replacement.
  - Road movement and clearances.
  - Casualty reporting and accountability.
  - Patient-holding procedures.
  - Force protection.
  - Convoy operations.
  - Geneva Conventions.
  - Detainees/EPW.
  - Improvised explosive devices (IEDs).
  - Religious and cultural considerations.

RECONDITIONING CENTERS

3-31. The medical company, CSC task-organizes CSC elements to staff separate small reconditioning centers in locations that are relatively secure. However, under some circumstances, the company may consolidate teams to establish a large reconditioning center, which supports two or three divisions. Reconditioning facilities normally locate near a CSH. See Chapter 12 for definitive information on reconditioning center operations.

Chapter 4
Unit Needs Assessment

SECTION I — PRINCIPLES AND TENETS OF UNIT NEEDS ASSESSMENT

GENERAL PRINCIPLES

4-1. The UNA is a systematic process for identifying the COSC needs of units. The UNA allows COSC personnel to identify priorities for interventions and activities and for allocating resources. The UNA is not a clinical screening to identify individuals who have or are at risk for BH disorders problems, but rather evaluates the needs of the Soldier population and leads to more effective preventive COSC activities and early interventions. The UNA allows COSC personnel to—

- Identify and describe specific areas of COSC need.
- Discover factors contributing to the needs.
- Provide an assessment of the BH training needs of Soldiers, leaders, UMT, and medical personnel within the unit.
- Develop plans to meet or improve the COSC needs of Soldiers and units through prevention and early intervention activities.

SCOPE

4-2. Unit needs assessments can be conducted at various command levels from small to large units. The COSC personnel at each level should conduct UNAs for their supported units. Larger-unit UNAs can include the composite findings and recommendations of one or more UNAs completed at subordinate levels. For example, a brigade UNA may include the results of several subordinate battalion UNAs (consolidated for confidentiality). Generally, UNAs are not conducted below the company level, though exceptional circumstances may dictate a platoon or lower UNA.

TENETS

4-3. Tenets of UNAs—

- Gain commander approval and support prior to conducting the UNA.
- Protect anonymity and ensure confidentiality of Soldiers and commanders. This includes the protection of unit identification from higher headquarters.
- Provide the commander with an unbiased assessment.
- Consider the social, political, and organizational factors of the environment.
- Ensure that information sources represent the entire unit.
- Select an assessment method that is consistent with the operational situation.
- Limit overgeneralizing the findings from one unit to another or from one time or situation to another.
- Recognize that UNAs provide population-level assessments of COSC needs, not clinical screening tools to identify individuals who may benefit from COSC interventions.
- Distinguish between what respondents report they need and what interventions are required.
- Ensure the UNA is planned and coordinated at a level commensurate with the complexity of the assessment and/or situation before starting.

SECTION II — FOCUS AND METHODS OF DETERMINING UNIT NEEDS ASSESSMENT

AREAS OF FOCUS FOR A UNIT NEEDS ASSESSMENT

4-4. A UNA involves the systematic assessment of numerous areas of Soldier and unit functioning. A typical UNA includes, but is not limited to, the following areas:

- Major stressors impacting the unit.
- Level of unit cohesion.
- Well-being of unit Soldiers.
- Soldier concern about home-front issues.
- Soldier knowledge and skill for controlling combat and operational stress.
- Soldier ideas for addressing COSC needs.
- Soldier knowledge of accessing COSC resources.
- Barriers and stigma that prevent Soldiers from accessing COSC services.
- Training needs of Soldiers, leaders, UMT, and medical personnel on topics of COSC importance (such as buddy aid, suicide awareness, or suicide prevention). See FM 6-22.5, FM 22-51, and US Army Training and Doctrine Command (TRADOC) Pamphlet 600-22 (available at: http://www.tradoc.army.mil/tpubs/pamndx.htm) for additional information on suicide awareness and prevention.

METHODS OF UNIT NEEDS ASSESSMENTS

4-5. The UNA takes advantage of all available information. Various methods can be used to assess general unit needs and to identify issues that differ among subgroups (for example, gender, rank, or race/ethnicity). The use of multiple assessment methods is recommended. These methods may include—

- Interviewing Soldiers to hear their perceptions and concerns. Information gathered during the casual conversation, although informal, may afford valuable anecdotal information (as well as developing trust and familiarity).
- Reviewing policy documents (standing operating procedures [SOPs]).
- Interviewing key unit personnel (chain of command, chaplain, and medical personnel).
- Conducting structured group interviews (focus groups or unit survey interviews).
- Administering standardized surveys and questionnaires (paper/pencil or web-based).
- Monitoring trend indicators (such as high rates of BH referrals, sick call, or misconduct, Soldier suicide, sexual assault, fratricide or disciplinary actions).
- Using multiple methods for gathering information when possible to ensure different viewpoints are considered.

Planning Considerations for Unit Needs Assessment

4-6. The UNA varies in complexity and formality depending on the purpose of the assessment and the needs of the supported unit commander. Many factors determine the complexity and formality of a UNA, which in turn influence its feasibility.

Complexity

4-7. Complexity equates to the requirements and cost to complete a UNA. Complexity is influenced by factors such as—

- The size and number of units to be assessed.
- Geographic dispersion of the units and time constraints.

Formality

4-8. Formality equates to the degree that scientific principles and methods are employed in the conduct of the UNA. An example of a UNA with low formality is one where the assessing team uses nonstructured group and individual interviews and perhaps brief questionnaires they have developed. The information is obtained from key individuals and a convenience sample of troops. That may be sufficient to quickly identify problems and make recommendations to resolve them. A UNA with high formality is one that uses professionally validated questionnaires and structured focus group interviews according to standard protocols. The data is collected from a scientifically selected sample of the larger population. The data is analyzed by standard analysis programs, so that statistically selected samples of valid comparisons can be made with other similarly sampled units in other geographical locations and across time. The OSAT expertise and assistance may be necessary for some UNA instruments and methods. In general, increasing formality increases the complexity of the UNA. The UNA requires a higher level of formality as the need for objective data and scientific precision increases (such as when results from multiple units are to be merged or compared with other UNAs).

Feasibility

4-9. Feasibility is the ability to accomplish a UNA with available resources. In developing a particular UNA, trade-offs are made to achieve an acceptable level of data quality (formality) for an affordable cost (complexity).

UNIT NEEDS ASSESSMENT PROCESS

4-10. All UNAs are conducted following a three phase plan.

PREASSESSMENT

4-11. Preassessment is an initial phase to obtain command support, determine target issues, and select appropriate methods to use.

ASSESSMENT

4-12. Assessment is the phase for gathering, integrating, and analyzing information to identify the COSC needs of the unit.

POSTASSESSMENT

4-13. The principal task of the postassessment phase is to determine the COAs to present to the commander which address the identified COSC needs. These findings are then linked to a plan of action.

WHEN TO CONDUCT UNIT NEEDS ASSESSMENTS

4-14. Unit needs assessments may be conducted at various times throughout the deployment cycle. A UNA may be conducted—

- Prior to initiation of COSC interventions and activities while COSC personnel and units establish their support relationships.
- To assess the effectiveness of COSC interventions and activities that are in the process of being conducted or that have been completed.
- At the request of supported commanders.
- To monitor trend indicators (see Paragraph 4-5).
- After serious traumatic events and significant unit transitions.
- To collect unit information for COSC planning.

4-15. Unit needs assessments should be conducted during all types of deployments, including stability and reconstruction operations, and combat operations.

Chapter 5

Consultation and Education

SECTION I — PRINCIPLES AND TENETS OF CONSULTATION AND EDUCATION

GENERAL PRINCIPLES

5-1. Combat and operational stress control consultation is defined as the transmission of information through an interactive relationship between the consultant and consultees. Education is used here in a broader sense in that it is the transmission of information by any means. Examples of consultation include providing COSC advice, coaching, training, and planning assistance. During this process, the consultant learns about the consultees and their needs, and tailors the interactions accordingly. Examples of education include distributing flyers, video and radio broadcasts, and news articles. The consultation and education functional area supports the other COSC functional areas.

CONSULTANTS

5-2. The term BH consultant is used in the general sense to describe any person performing COSC consultation or education. In addition to those theater or subordinate commander COSC consultants identified in Chapter 3, all BH/COSC personnel may serve as consultants at their level. Familiarization training with other BH disciplines enriches the ability to serve as a consultant.

CONSULTEES

5-3. Depending on the units in the area of support, broad types or categories of personnel may be consultees. These personnel are involved in recognition and control of stress as a result of their position or duty assignments. Consultation and education consultees may include—

- The command or unit surgeon and his staff.
- Staff chaplain and UMT.
- The senior commander and the senior NCO of a battalion, brigade, division, or corps.
- Staff officers and NCOs, including adjutant and personnel (S1/G1), intelligence (S2/G2), operations (S3/G3), civil-military affairs (G9), and the Judge Advocate General (JAG).
- Medical personnel such as PVNTMED teams that have missions that often complement the COSC mission.
- Company grade leaders, especially company commanders, executive officers, first sergeants (1SGs) platoon leaders, platoon sergeants, company NCOs, and Soldiers that are trained to be peer mentors.

SOLDIER-PEER MENTORS

5-4. Soldiers selected by their commanders may be trained to provide COSC help-in-place assistance for COSC information to peers. They may also serve as a POC between fellow Soldiers and the COSC and UMT teams. Selected Soldier-peer mentors with additional training could assist commanders with conducting COSC training in mission risk assessments.

TENETS FOR CONSULTATION AND EDUCATION

5-5. Consultation and education are ongoing processes that are performed across the deployment cycle and the continuum of operations. They may be provided in response to a specific request by command or recommendation of COSC personnel. Consultation and education may be provided during routine scheduled meetings such as a commander's weekly update or a leader professional development class.

5-6. Effective consultation is accomplished by the consultant's active outreach. Consultation is best conducted through recurring face-to-face contact, preferably at the consultee's location. Telephone and radio may be used to setup initial meetings and provide follow-up consultation. Consultation may be conducted one-on-one or in small groups where interaction is feasible. When necessary and feasible, audio or video teleconference may suffice. Active outreach supports the functional areas particularly COSC triage and traumatic event and transition management.

5-7. Successful consultation depends on the consultant's credibility, and the trust and familiarity established with consultees. In addition to sound professional knowledge base and clinical skills, the consultant must have military bearing and knowledge of the military (including the units, missions, vocabulary, acronyms, and skills involved). Rapport is enhanced by the demonstration of the consultant's genuine interest in the consultee and the unit. There are some situations when consultation and education are more effectively provided by the COSC officers, the NCOs, or the junior enlisted personnel.

SECTION II — CONSULTATION, EDUCATION, AND PLANNING

CONSULTATION PROCESS

5-8. The following six steps outline the process for most consultations:

- Initiate the process by introducing yourself and your capabilities to key leadership and get approval to continue.
- Assess the needs of consultees and formulate ways to address them.
- Present COA to consultee and define goals and feasibility of alternative actions.
- Implement approved COA.
- Evaluate outcomes or progress.
- Plan follow-up actions.

5-9. Consultation and education encompasses a broad range of topics, extending from prevention to treatment. The process of consultation and education assists the consultee with anticipation, identification, and control of stressors (environmental, physiological, cognitive, and emotional) and stress reactions (adaptive and maladaptive stress reactions). The following list of COSC functions, resources and their availability provides examples of consultation and education topics:

- Operation risk factors.
- Individual risk factors.
- Stress moderators.
- Unit cohesion, esprit de corps, and morale.
- Critical event management.
- Barriers to care and overcoming stigma.
- Operations planning.
- Integration of new personnel to unit.
- Trust in equipment and supporting units.
- Tough realistic training.
- Home-front stressors.
- Indicators of unit stress level.
- Combat and operational stress behaviors.
- Combat and operational stress reactions.

- Misconduct stress behaviors.
- Mental disorders.
- Suicidal and homicidal behaviors.
- Command referral processes.
- Scapegoating behaviors.
- Rumor control.
- Physical needs (such as sleep, nutrition, and hygiene).
- Social/emotional needs (such as morale, welfare, recreation push package).
- Communication with home.
- Reintegration of Soldiers into unit.
- Alcohol and substance abuse.
- Procedures for COSC triage.
- Care provider stress (such as compassion fatigue, vicarious trauma, moral dilemma).
- Long-term consequences of COSR and mental disorders.
- Medical evacuation policy and procedures.
- Leadership.
- Fear management.
- Rest and relaxation (R&R) considerations.

5-10. Additional details about some of these topics may be found in FM 22-51.

5-11. From theater staff level down to the MH sections, BH personnel are involved in the consultation and planning process for COSC support. Their support may be either through providing situational updates, developing staff estimates, or being directly involved and assisting with the development of the FHP plan. Planning starts with mission analysis. The mission analysis is the first step of the military decision making in abbreviated planning for a time-constrained situation, see FM 5-0. A part of mission analysis is based on the commander's intent and guidance. A medical battalion headquarters staff may receive addition instructions from the MEDBDE headquarters staff. During the planning process, command surgeons and medical planners (S3/G3) may seek information from the BH/COSC consultant or task medical company/detachment, CSC for COSC estimates.

MISSION ANALYSIS

5-12. The COSC planning is an essential part of every medical estimate and operations plan. Ensure that the command surgeon involves the COSC consultant in all medical planning. Mission analysis includes—

- Assessing COSC capabilities (organic and attached assets with current status and location).
- Gathering information from the commander's guidance regarding the upcoming operations. (See the S2 for intelligence, the S3 for scheme of maneuver and forces involved; the S1 for casualty estimates, personnel status issues, and units supported, and the logistics staff officer [S4/G4] for logistical considerations). Surgeons provide disease and nonbattle injuries (DNBI) estimates and status of medical units. Unit needs assessments, organic medical personnel, and UMT can provide additional information.
- Developing the COSC estimate based on the gathered information. The COSC estimate helps medical planners to anticipate demands on medical and COSC resources and to prepare accordingly. See FM 3-0, FM 5-0, and FM 8-55 for methods to develop the COSC estimate.
- Assessing limitations (specify reasons that BH assets are not available).
- Identifying specified, implied and essential COSC task in the FHP annex of the OPORD.
- Providing COSC COA.
- Presenting estimate and COA to command surgeon. Define goals and feasibility.
- Implementing approved COA. Consult with implementing COSC personnel about estimate and operational details.

- Evaluating outcomes or progress.
- Providing estimate and COA for fragmentary order (FRAGOs) and planning follow-up actions.

REHEARSAL

5-13. To achieve optimal synchronization, the FHP plan is rehearsed as an integral part of the combined arms plan at the combined arms rehearsal. Medical leaders provide input to the over all plan and develop the concept for the FHP plan. During the decision-making/orders process, they identify critical events and synchronize the FHP plans. In addition to medical locations on the CSS overlay, these plans indicate the triggers for FHP events. At the rehearsals, leaders practice their synchronized plans that include FHP.

5-14. The CSS/FHP annex of the OPORD that includes map overlays is the conclusion of the medical planning efforts; the rehearsal is the culmination of the preparation phase for an operation. The medical platoon leader has the responsibility for rehearsing FHP operations. Rehearsals are done to achieve a common understanding and a picture of how the plan will be implemented. The following leads to a successful rehearsal—

- All plans must be completed prior to the rehearsal.
- The FHP portion of the battalion rehearsals should focus on the events that are critical to mission accomplishment. A successful rehearsal ensures explicit understanding by subordinate medical personnel of their individual missions; how their missions relate to each other; and how each mission relates to the commander's plan. It is important for all medical units to understand the complete FHP concept.
- Rehearsing key FHP actions allows participants to become familiar with the operation and to visualize the "triggers" which identify the circumstances and timing for friendly actions. This visual impression helps them understand both their environment and their relationship to other units during the operation. The repetition of critical medical tasks during the rehearsal helps leaders remember the sequence of key actions within the operation and when they are executed.
- The OPORD is issued through effective troop leading procedures.

SECTION III — TRANSITION MANAGEMENT AND SUPPORT IN THE DEPLOYMENT CYCLE

SUPPORTING TRANSITION AND PHASES OF A DEPLOYMENT

5-15. Transition and phases of a deployment cycle may have different and sometimes unique stressors. These stressors may be subjects for consultation and education. Needs assessment may be necessary to redistribute COSC assets and change the means for delivering COSC activities and interventions.

5-16. The COSC personnel conduct transition (change of command) workshops, especially after relief of commanders. These workshops are normally requested by the incoming commander. The purpose of these workshops is to—

- Facilitate staff discussion of what the staff sees as the unit and staff's strong points and the areas needing more work.
- Provide the new commander the opportunity to discuss his leadership style and his expectations and set priorities for the staff.

5-17. The change of OPTEMP may include—

- Going from deployment to combat (offensive or defensive operations).
- Conducting battle handover in place by turning the battle over to another unit.
- Going from high intensity combat to stability and reconstruction operations.

PREDEPLOYMENT SUPPORT

5-18. Medical support in predeployment is mainly the responsibility of MEDDAC personnel or of a US Army Reserve (USAR) mobilization site augmentation unit (which has no BH personnel assigned).

Division and brigade COSC BH personnel are involved with their unit's Soldiers who are being treated for mental disorders, and with general COSC prevention of the unit members and families. The CSC units that are not deploying may be tasked to support predeployment COSC activities and interventions. Combat and operational stress control screening of individuals during predeployment may include—

- Secondary BH screening for Soldiers who are referred by primary care providers that flagged positive on DD Form 2795 (Pre-Deployment Health Assessment).
- Health screening of medical and BH records.
- Fitness for duty evaluations according to AR 40-501.
- Evaluations and recommendations to leadership about Soldiers who are fit for duty overall, but should not be deployed at this phase of the operation for BH reasons.

5-19. Other predeployment functions may be provided on request. These functions may include—

- Briefings, consultation and education on deployment cycle stressors and how to cope, and on specific stressors that Soldiers may encounter in the AO.
- Unit assessments at request of commanders.

Mid-Tour Rest and Recuperation or Emergency Leave

5-20. Leaders must be aware and alert for Soldiers who exhibit a need for COSC screening prior to mid-tour R&R. Other Soldiers may only require some education briefings and handout cards on handling their transition from a hostile CZ to R&R environments. In some cases, prior to emergency leave, some screening for domestic violence risk may also be necessary in any potential high-risk Soldier. This COSC screening is done for ensuring the safety of the Soldier and of others. One protocol has unit commanders referring those Soldiers that have requested leave for the purpose of addressing marital, legal, or other highly charged difficulties at home, to the BH officer. The BH officer explores the situation and the Soldier's reactions to it and checks for history with the rear detachment of the MH section, MEDDAC/MEDCEN and FAP. If the BH officer finds the Soldier to be at high risk, the commander can deny leave. Another option is to escort the Soldier to home base. At home base, he is kept by his unit's rear detachment until a meeting is held under safe conditions with the Soldier, BH, FAP personnel, his family, and the others involved in the difficulty. At this meeting the situation is defused or resolved before the Soldier is permitted to leave the rear detachment area.

Redeployment

5-21. For Soldiers, redeployment is the process of getting orders to redeploy, embarking towards the demobilization site, or deploying to another out-of-the-theater mission. Screening of these individual Soldiers may be a requirement and have established criteria that should be used to ensure appropriate screening is conducted. The theater or specific units may require that all primary care providers who perform screening interview, postdeployment health assessment have a designated BH consultant. Available COSC/BH personnel provide secondary one-on-one BH screening for Soldiers who are referred by primary care providers because they flagged positive on DD Form 2795. Some Soldiers may require brief treatment or referral may be required as a result of the screening.

End-of-Tour Stress Management

5-22. End-of-tour stress management is essential to reduce mental problems with Soldiers returning to their home station and/or families. This also promotes Soldiers seeking help early when problems occur. Mental health/COSC personnel consult with command on activities during redeployment and for postredeployment. Issues may include—

- Common mental disorders and potential misconduct may result when Soldiers are relieved of the stress and the focus of dangerous missions and are looking towards and preparing to go home.
- Memorial services for those Soldiers who died during the deployment.
- Recognition of outstanding performance and the equitable awarding of decorations.
- Closure ceremonies.
- Homecoming-reunion education briefing and training.

5-23. The end-of-tour debriefings (EOTDs) are like leader-led after-action debriefings, except that the time period covered extends from predeployment through to this point in redeployment. The EOTD should be conducted while the unit is still in the theater or at some place where transportation home pauses for one or more days. For each phase of the operation, the participants review what were the significant events and the problems or stressor that bothered the group (including critical events, if any). Equally important the group recalls what went well, how problems and stressors were overcome (at least by some members), and what positive memories and feelings they had and will take home.

Facilitators

5-24. The COSC or UMT facilitators should be invited to advise the leader and sit in if there are likely to be many negative memories and feelings that have disrupted or threaten unit cohesion. Advise leaders and troops on the importance of sustaining unit identity and contact with teammates beyond postdeployment and in the future.

Postdeployment

5-25. Postdeployment covers that period of time from embarking from theater, through scheduled postdeployment activities, and return to work after block leave. As with predeployment, medical support in postdeployment is mainly the responsibility of installation MTF, often augmented by a USAR mobilization site augmentation unit. Division and brigade COSC/BH personnel and CSC units that are themselves demobilizing as their units are deactivating will have limited involvement with the BH and COSC needs of fellow Soldiers. These personnel continue to provide, as needed, consultation to the command and to key personnel of the unit that they were assigned/attached. They consult with the installation MTF about BH and coordinate COSC needs that should be met. The CSC units that were not deployed may be tasked to support postdeployment COSC activities and interventions. The COSC personnel who are not demobilizing provide reintegration education. Screening of individuals postdeployment will continue and secondary screening will continue in postdeployment and be recorded on a DD Form 2796 (Post-Deployment Health Assessment). Each primary screener should have a designated BH consultant from the MEDCEN/MEDDAC. Secondary screening and treatment of referrals are provided by the installation MTF. Recent research with a returning division indicated that a secondary postdeployment screening from 90 and 120 days after returning from the theater may be necessary. This seems to be the time when persisting or delayed symptoms, problems, and perhaps impairment show need for intervention. All Soldiers should be advised about follow-up BH support resources. Soldiers that are leaving active service should also be informed about deployment-related entitlements and benefits.

Readiness Phase of the Deployment Cycle

5-26. Division/brigade MH sections and medical units, CSC, of Active Army and Reserve Component (RC) must use their training opportunities to the fullest. During these training exercises, they should provide COSC consultation, education, and training to supported units the same as they would in an actual deployment. Mental health personnel will perform many of their critical mission functions and should not just simulate them. The management of real COSR occurs in field and garrison/home-station settings. Triage and emergency stabilization may be required. Restoration may be appropriate in some field exercises. Even reconditioning can sometimes be provided as a multiday "course" at Active Army posts where CSC units are stationed. Further, TEM and transition management are also potential major roles for organic MH sections and medical units, CSC in peacetime. The RC as well as Active Army medical units, CSC have been used extensively to respond to traumatic events and to assist units that are in the premobilization, mobilization, or postdeployment phases of their deployment cycles.

Chapter 6
Traumatic Event Management

OVERVIEW OF TRAUMATIC EVENT MANAGEMENT

TAILORING TRAUMATIC EVENT MANAGEMENT

6-1. Traumatic event management blends other COSC functional areas to create a flexible set of interventions specifically focused on stress management for units and Soldiers following potentially traumatizing event (PTE). Like other functional areas, COSC providers must tailor COSC and TEM support to the needs of the unit and Soldier.

6-2. An event is considered *potentially traumatic* when it causes individuals or groups to experience intense feelings of terror, horror, helplessness, and/or hopelessness. Guilt, anger, sadness, and dislocation of *world view* or faith are potential emotional/cognitive responses to PTEs. Studies of Soldiers in Operation Iraqi Freedom (OIF) and Operations Enduring Freedom (OEF) have shown a correlation between exposure to combat experiences and BH disorders, most particularly acute stress disorder and PTSD. The following events should be monitored as PTEs for Soldiers and units:

- Heavy or continuous combat operations.
- Death of unit members due to enemy or friendly fire.
- Accidents.
- Serious injury.
- Suicide/homicide.
- Environmental devastation/human suffering.
- Significant home-front issues.
- Operations resulting in the death of civilians or combatants.

Scope of Traumatic Events Management

6-3. In this chapter, TEM is discussed as it applies to military units and personnel to support readiness. Traumatic events management can be adapted to nonmilitary groups or individuals brought together by a PTE. Traumatic events debriefing should be conceptualized as an ongoing process and not an acute intervention. For military units and personnel, TEM is active in all phases of the deployment cycle and across the continuum of military operations.

Traumatic Events Management Functional Mission Area

6-4. Successful TEM relies on a solid foundation of other COSC functional area activities such as UNA that may require the COSC provider to—

- Establish a credible working relationship with supported unit leadership.
- Understand the unit's needs via previously conducted UNAs.

6-5. In the absence of preexisting relationships and UNA, COSC providers should secure command support and recommend a UNA as the first step in TEM. Subsequent UNAs will clarify unit responses to other TEM interventions and to ongoing unit needs.

Unit Needs Assessments

6-6. When conducting a UNA after a PTE the COSC providers need to take under consideration the following—

- Ensure a timely arrival that does not disrupt unit operations but facilitates the UNA.

- They should not interrupt or intrude on the people who are attending to the acute crisis when arriving at the unit, unless asked. Presence without intruding will gain trust information and POC.
- Understanding that the UNA may be limited by the urgency of the unit's return to action, the difficulty of data collection, and having limited resources. Sufficient knowledge must be gained to tailor the interventions to the unit before initiating them. It is better to defer the intervention to the next opportunity if the unit must return to action immediately after replenishment and/or before necessary data is collected.
- Know that the UNA is a unit-level assessment and does not substitute for individual-level screenings or COSC triage. See Chapter 4 for a detailed discussion of UNA.

Consultation and Education

6-7. The COSC provider should conduct unit leader consultation and education activities prior to a PTE. Consultation and education topics should include—

- The impact of PTE on unit and Soldier readiness.
- Common PTEs for units and Soldiers.
- An overview of TEM.
- Components of TEM.
- Normal responses to PTEs.
- Triggers to refer Soldiers for BH evaluation.
- Development of SOPs for responding to PTEs when they occur.

6-8. Preemptive consultation and education prepare unit leaders to institute TEM interventions following a PTE. The 5 R's are a good model to build on. In acute TEM interventions, leaders should consider interventions that target—

- Safety, security, and survival.
- Food, hydration, clothing, and shelter.
- Sleep.
- Medication (replace medications destroyed or lost).
- Orientation of unit/Soldiers to developing situation.
- Restoration of communication with unit, dependents, friends, and community.
- Location should be at a secure site that provides protection from ongoing threats, toxins, and harm.

6-9. Consultation and education to Soldiers should emphasize normalizing the common reactions to trauma, improving their coping skills, enhancing self-care, facilitating recognition of significant problems, and increasing knowledge of and access to COSC services. Post-PTE UNAs guide further consultation and education efforts. Combat and operational stress control providers should be aware that leaders may not have experienced the PTE in person, but have experienced the PTE through their Soldiers and may require support.

6-10. For an in-depth discussion of consultation and education, refer to Chapter 5.

Combat and Operational Stress Control Triage

6-11. The COSC provider should be prepared to provide COSC triage in the aftermath of the PTE. Consultation and education is important to ensure appropriate and timely referrals for triage. The COSC provider can offer the following guidelines in referral of Soldiers for COSC triage:

- Persistent or worsening traumatic stress reactions (such as dissociation, panic, autonomic arousal, and cognitive impairment).
- Significant functional impairments (such as role/work relationships).
- Dangerousness (suicidal or violent ideation, plan, and/or intent).

- Severe psychiatric comorbidity (such as psychotic spectrum disorder, substance use disorder, or abuse).
- Maladaptive coping strategies (such as pattern of impulsivity or social withdrawal under stress).
- New or evolving psychosocial stressors.
- Poor social support.
- Failure to respond to acute supportive interventions.
- Exacerbation of preexisting psychiatric conditions.
- Soldier request for assessment.

Stabilization

6-12. The COSC provider should be prepared to provide or coordinate stabilization services following the PTE. Pre-PTE coordination with medical unit personnel promotes safe management.

Soldier Restoration

6-13. The measures below are applicable to Soldiers with COSR following a PTE. The COSC provider should be familiar with the 5 R's and with BICEPS. In keeping with restorative efforts, the COSC provider focuses on the following measures through leadership consultation, Soldier education, and/or direct management:

- Minimizing exposure of Soldiers with COSR to further PTE.
- Reducing physiological arousal.
- Mobilizing support for those who are most distressed.
- Providing information and fostering communication and education.
- Using effective risk communication techniques.
- Proving assurance/reassurance.
- Mitigating fear and anxiety.
- Encouraging sleep hygiene.
- Reestablishing routines.
- Promoting exercise and nutrition.
- Encouraging self-paced emotional ventilation.
- Discouraging use of alcohol/substances.

Behavioral Health Treatment

6-14. Given the correlation between PTE and development of traumatic stress disorders the COSC provider must be familiar with the best current practices for evaluation and treatment. The Veterans Health Administration (VHA)/DOD Clinical Practice Guidelines website (http://www.oqp.med.va.gov/cpg/cpg.htm) offers clinicians evidence-based assessment and treatment algorithms for acute stress disorder, PTSD, and many other BH disorders.

6-15. In recent years, the use of early interventions in response to PTE has come under critical review. The focus of much of this debate is on the use of psychological debriefing (PD) and more specifically, the critical incident stress debriefings provided to individuals or groups exposed to PTEs. Current research suggests that PD/critical incident stress debriefings can be harmful to participants, while failing to reduce traumatic stress reactions or to prevent the progression to PTSD. As PD/critical incident stress debriefings undergo definitive study, COSC providers should adhere to current evidence-based best practices.

6-16. In accordance with the VHA/DOD Clinical Practice Guidelines (http://www.oqp.med.va.gov/cpg/cpg.htm) for PTSD, COSC providers should—

- Consider alternative methods to PDs for individuals affected by PTEs.
- Avoid PD as a means to reduce acute posttraumatic distress (acute stress reaction or acute stress disorder) or to slow progression to PTSD.

- Understand there is insufficient evidence to recommend for or against conducting structured group debriefings.
- Be aware that compulsory repetition of traumatic experiences in a group may be counter-productive.
- Consider group debriefings with preexisting groups (such as teams, units, emergency medical treatment [EMT] teams, coworkers, family members) may assist with group cohesion, morale, and other important variables that have not been demonstrated empirically.
- Emphasize that group participation must be voluntary.

LEADER-LED AFTER-ACTION DEBRIEFING

6-17. A leader-led after-action debriefing is lead by a platoon, squad, or team leader and is not normally conducted above platoon level. The leader-led after-action debriefing should be conducted after all missions especially when the maneuvers did not go according to plan. A leader-led after-action debriefing may even be sufficient for PTEs involving injury or death. For the leader to conduct a PD, his personnel should have received previous PDs; and normally provide peer support and validation for showing and talking about their emotional reactions during and after the debriefing. The best time to conduct this debriefing is as soon as is feasible after the team/squad/platoon has returned to a relatively safe place and members have replenished bodily needs and are no longer in a high state of arousal. Usually a well-conducted leader-led after-action debriefing is the best option to manage PTEs during a mission. The exception to this type of debriefing is when the event evoked reactions that seriously threaten unit cohesion and/or have a high likelihood of arousing disruptive behavior and emotions. In these situations the leader should ask himself the following—

- Should I conduct the debriefing?
- Should a trained facilitator be present?
- Should a request for COSC TEM be submitted for his team/squad/platoon?

CONDUCTING A LEADER-LED AFTER-ACTION DEBRIEFING

6-18. These debriefings require the leader to extend the lessons-learned orientation of the standard AAR. He uses the event reconstruction approach or has the individuals present their own roles and perceptions of the event, whichever best fits the situation and time available. See Training Circular (TC) 25-20 for definitive information on AARs. When individuals express or show emotions, the leader and the teammates recognize and normalize them; they agree to talk about them later and support the distressed Soldier through personal interactions. The group then returns to determining the facts. Lessons-learned discussion is deferred until all the facts are laid out. See FM 22-51 for additional information. The leader may provide education about controlling likely reactions or referral information at the end, depending on his knowledge and experience.

6-19. When a PTE is likely to create individual or collective guilt, distrust, or anger, the unit leader should be encouraged to request COSC assistance. Either a COSC or a UMT person trained in TEM sits in with the leader-led debriefing as a familiar and trusted friend of the unit. The COSC or UMT facilitator helps the unit/team leader rehearse and mentors the leaders on the debriefing process. During the PD, the facilitator can ask questions of the group to clarify the facts and steer the process away from divisive anger, blaming, and scapegoating toward a positive, cohesion-restoring outcome. This method is halfway between a simple leader-led after-action debriefing and a PD and is referred to as a *facilitated leader-led after-action debriefing*. The leader conducting the debriefing must be attentive to identify individuals needing COSC follow-up.

6-20. Leaders in positions above platoon level also have a role in leader-led after-action debriefing. Company commanders and 1SGs may conduct leader-led after-action debriefing with their subordinate leaders. Battalion commanders and higher may conduct leader-led after-action debriefings with their staff after distressing actions and may include subordinate leaders when time allows bringing them together.

COOL DOWN MEETINGS

6-21. A cool down meeting is referred to as an immediate, short meeting when a team or larger unit/group returns from the battlefield or other missions. These cool down meetings are held after heavy/intense battles with the enemy or a shift in the mission has occurred which is highly arousing and/or distressing. This is especially important after critical events. The cool down meeting is an informal event and occurs before the participants fully replenish their bodily needs and precedes any other activities including more COSC interventions, or return to the mission.

6-22. Personnel who coordinate and wait for the return of the unit that has been in a heavy intense battle, include leaders or supporting officers or NCOs from the command, UMT, and COSC providers. (In domestic support operations the same personnel identified above or other trained personnel from governmental or nongovernmental organizations such as the Red Cross may be waiting on the return of the unit after a PTE. These personnel may be present at the cool down meeting.)

COMPONENTS OF A COOL DOWN MEETING

6-23. Components of a cool down meeting may include—

- Assembling all of the unit personnel at a safe and relatively comfortable location for a brief period of time (about 15 minutes).
- Receiving or sharing nonstimulating beverages and convenient food (*comfort* foods if available).
- Providing personnel the opportunity to talk among themselves.
- Giving recognition and praise for the difficult mission they have completed.
- Providing information to unit personnel on where and how they will rest and replenish.
- Previewing the immediate agenda for the unit on what will happen after the cool down meeting including plans for further debriefing and/or other available stress control or morale/welfare intervention.
- Providing announcements pertaining to further preparations and expected time of return to the mission.

6-24. The COSC personnel may have a role in cool down meetings. In consultation and education, COSC consultants emphasize the value of cool down meetings and the simplicity of the components, which are easily neglected in crisis situations. When feasible, they unobtrusively attend the cool down meeting, showing "presence" while learning of the event, getting familiar with the key people, and observing anyone showing signs of distress and being available to them. If requested, they may give a very brief introduction and review of normal stress reactions that unit personnel may have in the next few hours.

Chapter 7

Combat and Operational Stress Control Support for Units Undergoing Reconstitution

UNIT RECONSTITUTION SUPPORT

7-1. Unit reconstitution support is defined in FM 100-9 as extraordinary actions that commanders plan and implement to restore units to a desired level of combat effectiveness commensurate with mission requirements and available resources. Besides normal support actions, reconstitution may include—

- Removing the unit from combat.
- Assisting the unit with external assets.
- Reestablishing the chain of command.
- Training the unit for future operations.
- Reestablishing unit cohesion.

7-2. When tasked as part of a reconstitution task force, COSC personnel are responsible for providing units with Soldier restoration, performing the COSC functions of the UNA, and providing consultation and education, as required. They provide triage, stabilization, Soldier restorations, and short time-constrained COSC treatment, when needed. Reconstitution is a time-constrained process, but Soldier restoration may be provided at the reconstitution site if several days are available. The focus of consultation and education efforts include—

- Rebuilding unit cohesion.
- Integrating new Soldiers or groups of Soldiers into the unit.
- Facilitating assumptions of command by replaced leaders.
- Facilitating the building of Soldiers' confidence in their leaders and themselves.
- Mentoring unit leaders in leader-led after-action debriefing process.
- Advising on COSC aspects of bereavement memorial services and communication with the family support group and unit families. The COSC personnel also facilitate leader-led after-action debriefings and lead or serve as observers in TEM interventions.

RECONSTITUTION PROCESS

7-3. Reconstitution of units transcends normal day-to-day force sustainment actions. Reconstitution is a total process. Its major elements are reorganization, assessment, and regeneration, in that order. All COSC personnel should be thoroughly familiar with FM 100-9.

Reorganization

7-4. Unit reorganization primarily involves a shifting of internal resources and is accomplished as either immediate or deliberate reorganization. The commander of the attrited unit decides to reorganize when required and may consider the following—

- Immediate reorganization is the quick and usually temporary restoring of degraded units to minimum levels of effectiveness. Normally, the commander implements immediate reorganization at his combat location or as close to that site as possible to meet near-term needs. The COSC personnel provide consultation with key POC in the unit (leaders, UMTs, and trauma/health care specialist, usually by telecommunication).
- Deliberate reorganization is done to restore a unit to the specified degree of combat effectiveness. Usually, more time and resources are available further to the rear. Procedures are similar to immediate reorganization except that some personnel and weapons system

replacement resources may be available, equipment repair is more intensive, and more extensive cross- leveling is possible. When used in reorganization, cross-leveling involves the movement of personnel and/or equipment between units to achieve equalization. The process is accomplished while maintaining or restoring the combat effectiveness of the units involved.

- The role of COSC personnel in deliberate reorganization may require a COSC team to deploy to the reorganization site. They assess the stressors and stress reactions and advise the commanders on supportive actions, (such as those for the reorganization of small unit-level elements), assists command with transitions and integration of new replacements.
- When the reorganization involves battalion size or larger units the process becomes more like the coordinated operation described for unit regeneration, provided below. It is likely to draw upon fitness team expertise and Soldier restoration assets from rearward as well as on the forward deployed teams.

REGENERATION

ASSESSMENT PHASES

7-5. The defining characteristic of regeneration is the massive infusion of personnel, equipment, and assistance under the directions of higher headquarters. Assessment measures a unit's capability to perform its mission. It occurs in two phases. The unit commander conducts the first phase. He continually assesses his unit before, during, and after operations. If he determines it is no longer mission-capable even after reorganization, he notifies his commander. The higher headquarters either changes the mission of the unit to match its degraded capability or removes it from combat. External elements may also have to assess the unit after it disengages. This is the second phase. These elements do a more thorough evaluation to determine regeneration needs. They also consider the resources available.

7-6. The second phase of assessment begins with an initial survey by a team sent by the higher headquarters. This team determines the status and needs of the attrited and exhausted unit as it moves to the regeneration site. Some of the key issues in estimating the COSC needs of the unit include the—

- Percentage and nature of casualties.
- Duration of operations and environmental exposure.
- Status of hydration, nutrition, and sleep.
- Loss and current effectiveness of leaders.
- Attitudes, perceptions, and level of confidence of unit survivors.
- Traumatic events.

7-7. The COSC consultants at every level must provide consultation to all command surgeons on the importance of including BH/COSC personnel in all reconstitution operations. The initial evaluation team should include a subordinate commander COSC consultant. Task-organized CSC teams normally deploy to the unit during the second phase to provide UNA, consultation, and other COSC activities and interventions. Unit needs assessment with high formality may be requested by the higher headquarters for decision-making purposes. These assessments are feasible for selected units, with prior coordination, as much of the logistical complexity is reduced by the orchestrated planning at higher command echelons.

Conducting Regeneration

7-8. Regeneration involves rebuilding a unit through large-scale replacement of personnel, equipment, and supplies. If required, it includes internally reorganizing; reestablishing or replacing essential C2 and conducting mission-essential training for the reconstituted unit. Regeneration is required when heavy losses of personnel and equipment leave a unit combat-ineffective and unable to continue its mission. The commander controlling assets to conduct regeneration decides whether to use resources for this purpose. Regeneration has two variations: incremental regeneration or whole-unit regeneration. Incremental regeneration is the massive infusion of individual personnel replacements and single items of equipment into the surviving unit elements. Whole-unit regeneration is the replacement of whole units or definable subelements, such as squads, crews, and teams. The S1/G1, S4/G4, and medical staffs coordinate the dispatch of the regeneration task-force teams. These teams occupy the reconstitution site before arrival of the exhausted unit. The COSC assets that perform the indicated initial assessment are needed as part of this task force. The reconstitution task force guides each element of the arriving units into its designated areas. The regeneration task force provides for the immediate needs of the survivors. This should include personal gear and equipment to replace lost or damaged items, food services, personnel services, maintenance teams, and medical teams to provide sick call services while organic medical personnel rest. Replacement personnel are sent to the reconstitution site. The COSC personnel assist with their assimilation into the regenerated unit.

RECONSTITUTION RESOURCE REQUIREMENTS FOR COMBAT AND OPERATIONAL STRESS CONTROL

7-9. Factors which influence the resource requirements for COSC in reconstitution (deliberate reorganization and regeneration) include the size of the unit, number of subunits which have suffered heavy casualties, the extent of emotional trauma, and time available. The focus of COSC for reconstitution support was provided in Paragraph 7-2. A guideline is provided in Table 7-1 for COSC personnel requirements. This staffing should not overly rely on organic COSC staff that will be in need of rest.

Table 7-1. Reconstitution operations guideline for combat and operational stress control personnel requirements

Size of Unit	*Personnel Required*
Company/Troop	2 to 4
Battalion/Squadron	6 to 12
BCT/Brigade Regiment	12 to 30
Division	30 to 60

Chapter 8

Combat and Operational Stress Control Triage

SECTION I — THE TRIAGE PROCESS FOR COMBAT AND OPERATIONAL STRESS CONTROL

SOLDIER TRIAGE

8-1. The COSC triage process is the sorting of Soldiers based on an assessment of their needs and capabilities, and the location where they can best be managed in keeping with BICEPS principles. Triage is applicable at every level of care. The two key components of COSC triage are assessment and disposition.

8-2. Assessment is an evaluation of the Soldier's physical and BH needs, potential medical emergencies, and other safety risks. Assessment is performed by COSC personnel according to their professional training, expertise, and standards.

8-3. Disposition is the COSC intervention plan to address the needs identified in the assessment. Disposition has two components that include—

- Determining what intervention techniques best address the Soldier's needs and functional capabilities.
- Selecting the best location where the Soldier can be managed. The personnel conducting COSC triage should consider the needs, abilities, and the safety of the Soldier. They should also consider the unit's capacity to provide COSC interventions based on its OPTEMPO mission, resources, response to prior consultations, and willingness to participate in COSC interventions.

TRIAGE ALGORITHM FOR COMBAT AND OPERATIONAL STRESS CONTROL

8-4. Like surgical triage categories, COSR also uses triage categories. Each of the COSR triage categories are discussed in detail in the below paragraphs. The COSC triage algorithm presented in Table 8-1 uses some of the triage categories.

Table 8-1. The combat and operational stress control triage algorithm

<table>
<tr><td>Step 1</td><td colspan="5">Is this a medical emergency?</td></tr>
<tr><td></td><td>Yes</td><td>Refer to nearest MTF</td><td>No</td><td colspan="2">Go to Step 2</td></tr>
<tr><td>Step 2</td><td colspan="5">Does the Soldier require medical/behavioral observation?</td></tr>
<tr><td rowspan="3"></td><td rowspan="3">Yes</td><td rowspan="3">Go to Step 3 A</td><td rowspan="3">No</td><td colspan="2">Does Soldier have presumptive COSR or MH disorder?</td></tr>
<tr><td>Yes</td><td>Go to Step 3 B</td></tr>
<tr><td>No</td><td>Help-in-place</td></tr>
<tr><td>Step 3</td><td colspan="2">A. Can the Level II MTF or CSC Soldier Restoration Center provide adequate evaluation and intervention?</td><td colspan="3">B. Can the Soldier's unit support the 5 R's or other treatment interventions?</td></tr>
<tr><td rowspan="4"></td><td>Yes</td><td>HOLD*</td><td>Yes</td><td colspan="2">UNIT</td></tr>
<tr><td rowspan="3">No</td><td rowspan="3">REFER</td><td rowspan="3">No</td><td colspan="2">Is there a suitable support unit?</td></tr>
<tr><td>Yes</td><td>REST</td></tr>
<tr><td>No</td><td>HOLD*</td></tr>
<tr><td colspan="6">* When deciding between two or more potential Level II MTFs or CSC Restoration Centers, refer the Soldier to the one closest to his unit that meets his COSC needs.</td></tr>
<tr><td>Step 4</td><td colspan="5">Use on subsequent triages. Has the Soldier improved after appropriate duration of intervention?</td></tr>
</table>

TRIAGE CATEGORIES FOR COMBAT AND OPERATIONAL STRESS REACTION CASES

8-5. The following are triage categories that may be used for COSR cases. *Help-in-place* (HIP), *rest*, *hold*, and *refer* cases are discussed below.

HELP-IN-PLACE CASES

8-6. *Help-in-place* is used to identify those cases that do not have severe COSR or BH disorders. They are provided COSC consultation and education, as appropriate, and remain on duty. These interactions may occur in any setting (for example, dining facility, workplace, or the post exchange). Individual identifying information is not retained or documented. There is no implicit or explicit therapist-patient or therapist-client relationship in HIP interactions.

8-7. The unit identifies those cases that remain with or return to their original unit, either for full duty with their section/platoon or for light duty with extra rest and replenishment within a headquarters element. This option depends on the unit's mission, resources, and the Soldiers symptoms. Personnel performing triage must, therefore, be familiar with the unit's situation and take that into account. When the Soldier's condition improves, the Soldier and/or unit may not feel that additional triage is necessary.

Note. *Help-in-place* is a new triage category that replaces the old *duty* category and is also used for unit cases.

REST CASES

8-8. *Rest* identifies those cases that are provided rest and replenishment in a nonmedical support unit, usually one that is in support of their unit. These Soldiers do not require close medical or BH observation or treatment. They are unable to return immediately to their own unit either because their unit cannot provide an adequate environment for the 5 R's; or transportation is not available for at least a day; or the 5 R's can best be coordinated from the nonmedical support unit. This option depends on the resources and mission of the available CS/CSS units, as well as on the Soldier's symptoms. Someone must be designated to be in charge of the Soldier and ensure that the 5 R's are provided. There must be a reliable transportation link to return the Soldier to his original unit after a day or two. When the Soldier's condition improves sufficiently for him to return to his unit, the Soldier and/or the supporting unit may feel that additional triage is necessary.

HOLD CASES

8-9. *Hold* refers to those cases that require close medical/BH observation and evaluation at a Level II MTF or COSC Soldier restoration center because the Soldiers symptoms are potentially too disruptive or burdensome for any available CSS unit or element. Soldier's symptoms may be caused by a BH disorder that could suddenly turn worse and require emergency treatment. The Level II MTF or COSC Soldier restoration center must have the capability to provide the necessary medical observation, diagnostic tools, and adequate stabilization for emergency treatment. When deciding among capable Level II MTFs or COSC Soldier restoration centers, refer the Soldier to the one closest to his unit that meets his COSC needs. Assessment of closeness considers speed and reliability of transportation and back. Consider transferring to another Level II MTF or COSC Soldier restoration center with increased capabilities before changing a Soldier's triage category to *refer*. All *hold* cases will be triaged again by COSC personnel or other trained medical personnel after they have been placed in this category.

REFER CASES

8-10. *Refer* cases are similar to the *hold* cases, except that *refer* cases are too disruptive and burdensome for the MTF or the COSC Soldier restoration center that is not resourced to care for this particular case. The MTF or COSC Soldier restoration center cannot provide the necessary level of diagnostic and treatment capabilities. *Refer* cases requiring care at a COSC reconditioning center, a Level III MTF or higher levels of care will be triaged by COSC or other trained medical personnel prior to being transferred to these facilities.

TRIAGE PERSONNEL FOR COMBAT AND OPERATIONAL STRESS CONTROL

8-11. All COSC personnel participate in the triage process according to their professional training, experience, and standards. Familiarization training among BH disciplines extends the effectiveness of all COSC personnel in triage skills. Medical care providers must be mentored to use the COSC triage process. A commander providing the 5 R's and placing a Soldier in a support unit for a temporary break does not equate to COSC triage.

TRIAGE CONSIDERATIONS

8-12. Triage should be initiated when the—

- Soldier is a self-referral.
- Chaplain has referred the Soldier.
- Medical personnel have requested a COSC consultation and referred the Soldier.

- Unit member/buddy has referred the Soldier.
- Leader has requested an informal referral.
- Soldier is a command-directed referral (see DODD 6490.1 and DOD Instruction [DODI] 6490.4).
- Combat operational stress control personnel observe a Soldier's behaviors which indicate possible COSR or a BH disorder.

8-13. Factors that influence an assessment may vary in depth and duration due to several other factors. These factors may include the—

- Referral source.
- Nature of the complaint.
- Observed needs.
- Medical/psychiatric history.
- Availability of resources.
- Amount of privacy for conducting assessment.
- Operational Tempo.
- Environmental conditions.
- Professional training of the person making the assessment.
- Command interest.
- Soldier cooperation.

Note. Regardless of these factors, COSC personnel are responsible for conducting the assessment in a timely manner within professional standards of practice.

DOCUMENTATION

8-14. An assessment is documented according to AR 40-66 whenever the Soldier—

- Is diagnosed with a BH disorder.
- Has a condition (or suspected condition) requiring emergency medical evaluation or treatment.
- Is prescribed medication.
- Is assessed to be a high safety risk (for example, homicidal or suicide, cognitive impairment, substance abuse, and impulsivity).
- Requests that documentation of his assessment be made in his medical records.
- Is evacuated beyond Level II MTF for further assessment or treatment.
- Is command-referred for a BH evaluation.

TRANSFER AND EVACUATION

8-15. All COSC personnel are responsible for knowing the transfer/evacuation policies and procedures within their AO. Policies and procedural information are available through the command surgeon, medical regulating officer (MRO) of the medical command, control, communications, computers, and intelligence (C4I) headquarters or major MTF. All relevant background and/or clinical documentation must accompany the Soldier during the transportation or evacuation process.

MODE OF TRANSPORTATION

8-16. Nonambulance transport is the preferred mode of transportation for COSR and nonurgent BH cases. Examples of nonambulance transportation include the Soldier's unit vehicles, supporting supply/logistics vehicles, and nonambulance medical vehicles. Ambulances convey patient status on Soldiers and often must be reserved for medical emergency. Under the provisions of the Geneva Conventions, ambulances must be used exclusively in the performance of humanitarian duties, therefore, they cannot be used to return Soldiers to duty and transport must be provided by the supported unit.

ESCORT

8-17. Depending on a Soldier's condition, an escort (either medical or nonmedical) may be necessary to provide safety, monitoring, and accountability during transportation or evacuation. The escort should be an NCO or officer of equal or greater rank/grade as that of the escorted Soldier. Escorts must be emotionally mature, responsible, and capable of conducting their escort duties. Frequently escorts carry the Soldier's clinical documentation to the destination MTF.

FEEDBACK

8-18. Good communication is essential for effective continuity of care during the transportation/evacuation process. The Soldier's unit must be informed about his location and status throughout the process. The sending party must provide sufficient documentation about the Soldier's condition, history, and administered interventions. The receiving party must provide feedback to the sending party regarding receipt of the Soldier and his documentation.

SECTION II — PRECAUTIONS AND DIFFERENTIAL DIAGNOSTIC PROBLEMS ASSOCIATED WITH COMBAT AND OPERATIONAL STRESS CONTROL TRIAGE

PRECAUTIONS FOR COMBAT AND OPERATIONAL STRESS CONTROL TRIAGE

8-19. Medical emergencies must be identified during COSC triage. Medical emergencies consist of physical illnesses or injuries and/or BH disorders that can result in permanent injury, disability, or death. Early identification of a medical emergency avoids unnecessary delay in definitive treatment. Medical emergencies can cause emotional and/or BH changes and may resemble COSR in presentation. The following conditions and behaviors could be medical emergencies. Therefore, it is important to ensure medical examinations and disposition of Soldiers that display the following conditions—

- Psychosis.
- Mania.
- Alcohol withdrawal.
- Substance intoxication.
- Delirium.
- Suicidal gesture, attempt, or high risk for suicidal behavior.
- Catatonia.
- Significant paresis, paralysis, and/or sensory loss.

8-20. Assessing for physical illnesses or injuries is a critical part of COSC triage. The COSC personnel must always consider physical illnesses or injuries that resemble COSR or BH disorders. Physical illnesses or injuries may not reach the threshold of a medical emergency, but must be recognized and appropriately treated. Assessing for physical illnesses or injuries requires an adequate review of body systems and a quick physical examination. The examination includes vital signs, examination of head, eyes, ears, nose, throat, chest, abdomen, and extremities with simple testing of reflexes and muscle strength. In field situations (Level I and II), negative or normal findings need to be documented on DD Form 1380 (US Field Medical Card). Refer to AR 40-66 for detailed information on patient accountability and management of individual health records. Any positive findings from the physical examination must be evaluated further. If the examiner has not checked various body systems, it is not reassuring to tell a Soldier that his physical or BH complaints are only COSR. All COSC personnel should receive familiarization training on basic medical examination techniques and in documenting medical information. Whenever a physical illness or injury is suspected, personnel should consult with their medical peers for further assistance. Some cases will require direct medical examination by a physician or physician assistant (PA). The COSC personnel should not order tests or procedures that do not directly influence case management. Medical tests may promote the patient role in the mind of the Soldier. Needless tests may delay a Soldier's RTD and encourage secondary gains.

DEFER DIAGNOSIS OF BEHAVIORAL DISORDERS

8-21. During assessment, COSC personnel must always consider BH disorders that resemble COSR, but defer making the diagnosis. The COSC personnel favor this default position to preserve the Soldier's expectations of normalcy (according to BICEPS). This is also done to avoid stigma associated with BH disorders and to prevent the Soldier identifying with a patient or sick role. Deferral is also preferred because some diagnoses require extensive history collection or documentation that is unavailable during deployment situations (such as personality disorders and attention deficit hyperactive disorder). It is possible that a Soldier can have a combination of COSR, BH disorders and physical illnesses/injuries at the same time. In such cases, COSC personnel must rely on their clinical experiences, training, and consultation with peers and medical personnel to distinguish among these sometimes overlapping conditions. Physical injuries/illnesses are treated at a MTF, however, the Soldier may return for further COSC interventions and activities. Deferral of diagnosis is preferred, but diagnosis can be considered if the Soldier—

- Presents for reemerging symptoms of a previously diagnosed and/or treated BH disorder.
- Presents for refill prescription of psychotropic medication.
- Has a medical condition or BH disorder listed above in paragraph 8-19.
- Is enrolled in a reconditioning program.
- Fails to improve after having received four to five days of continuous COSC interventions and activities at hold or refer status.
- Requires individual BH treatment.
- Is referred for multiple episodes of COSR.

DIAGNOSTIC CONSIDERATIONS FOR DIFFERENTIAL DIAGNOSTIC DISORDERS

Low-Grade Environmental or Stress-Related Illnesses

8-22. Low-grade environmental or stress-related DNBI illnesses can drain the Soldier's strength and confidence. For example, chronic diarrhea and slight fever may exhaust, demoralize, and contribute to COSR among Soldiers. These conditions should be treated medically, concurrently with physical replenishment, rest, reassurance and organized activities, which restore the Soldier's confidence. If they persist in spite of rest and symptomatic treatment, a more aggressive workup and treatment may be indicated.

Dehydration

8-23. Dehydration deserves special mention because it can be very insidious. Soldiers under battlefield or heavy work conditions become extremely dehydrated without feeling thirsty. This is especially likely in CBRN equipment, or in a desert/arctic environment. Refer to FM 4-25.10 for additional information on the prevention of dehydration and FM 4-25.11 for first aid measures.

Hyperthermia

8-24. Hyperthermia (overheating) in an otherwise healthy individual often first causes mild elation and excessive energy. This may be followed by irritability, disorientation, and confusion. When core body temperature climb above 106° Fahrenheit (F) or 41° Centigrade (C), the Soldier may become belligerent, combative, and have visual hallucinations. If brain temperature rises further, the Soldier collapses and convulses in heatstroke. Refer to FM 4-25.11 for first aid measures and FM 4-25.10 for additional information on prevention and first aid measures for heat injuries.

Hypothermia

8-25. Hypothermia may cause an individual to become disoriented when core body temperature falls below 95°F (34.6°C). The person may move and speak slowly. His skin looks and feels warm, leading him to take off clothing. He becomes disoriented, then unresponsive and may appear to be dead. Hypothermia is as likely in cool or even warm, wet climates as it is in extremely cold ones. Refer to FM 4-25-10 for additional information on prevention and FM 4-25.11 for first aid measures.

Overuse Syndromes

8-26. Overuse of muscles, joints, and bones that have not been prepared for the strain of field duties can lead to persisting stiffness, pain, swelling, and orthopedic injuries. If severe, these injuries may require evacuation to a hospital for evaluation. Even if these injuries are avoided, the unfit person who overexerts has days of stiffness, aching, and weakness. Such cases are likely to develop COSR if further demands are made on them.

Rhabdomyolysis

8-27. Rhabdomyolysis is one potentially dangerous complication of severe muscle overuse (and of heatstroke or crush injuries) in which myoglobin from damaged muscle cells injures the kidneys. It can cause fatigue, seizures, muscle tenderness, and muscle aches. A warning sign is dark (tea-colored) urine, but without laboratory testing, this is not easily distinguished from the concentrated urine of dehydration. Rhabdomyolysis is a medical emergency.

Head Trauma

8-28. Concussion may stun the individual and cause amnesia, residual confusion, and/or impulsive behavior. For any case of suspected head trauma or for any case of significant memory loss (especially for a discrete period of time), check scalp, eyes, ears, nose, cranial nerve signs, and vital signs for evidence of head injury. Cases with deteriorating mental status are medical emergencies. If one pupil becomes larger than the other, it is an extreme emergency requiring immediate hospitalization. Left untreated, the condition can progress rapidly to coma and respiratory arrest within hours. If a head injury is suspected, monitor mental status and vital signs periodically, especially respiration, even though physical findings are negative. Awaken the Soldier periodically to check mental status and pupil size (allowing sufficient time to recover from any grogginess on awakening). Continuous monitoring is appropriate if there are serious concerns about the risk.

Spinal Cord Trauma

8-29. Pressure, bruising, and hematomas of the spinal cord, as well as severing of the spinal cord, can cause spinal shock, with loss of sensory and/or motor functions below the level of the injury in the affected dermatome and muscle group patterns. The loss of function may be bilateral, unilateral, or partial. These cases could be confused with paralysis or sensory loss presentations of COSR. Further manipulation of a fractured spine can worsen or make permanent the spinal cord damage. Information from the history of onset, a cautious physical and neurological examination, or complete relief of symptoms following hypnosis or strong positive suggestions could demonstrate convincingly that this is only COSR. It is best to be cautious and keep the spine immobile during care and transportation.

Postconcussion Syndromes

8-30. Postconcussion syndromes may persist weeks to months beyond the period of acute concussion. Postconcussion syndromes may include perceptual or cognitive impairment, poor impulse control, and difficulty in planning ahead. These are often accompanied by cranial nerve deficits or soft neurological signs.

Abdominal Trauma

8-31. Ruptured spleen or other intraperitoneal bleeding may cause shock. The Soldier may arrive in a fetal position and be unresponsive but have reflex guarding due to peritonitis.

Air Emboli and Focal Brain Ischemia

8-32. High blast overpressures from incoming high explosive ordnance can produce air emboli (bubbles in the blood) and focal brain ischemia (small areas in the brain which cannot get oxygen because the blood flow has been interrupted). Nuclear explosions can do this, as can high explosives when shock waves are magnified by reflection within bunkers, buildings, and trenches. Ruptured eardrums, general trauma, and evidence of pulmonary damage should be detectable. Cases may have stroke symptoms (loss of muscle strength, loss of sensation in parts of the body, and/or speech disturbances), which may be subtle and mistaken for COSR.

Laser Eye Injury

8-33. Laser range finders/target designators cause small burns on the retina if they shine directly into the eye, even at great distances and especially if viewed through optics. If the laser beam causes a small retinal blood vessel to bleed inside the eyeball, the person will see red. If blood inside the eye is confirmed on examination, the Soldier should be evacuated to a hospital with verbal reassurance that he may RTD soon. If the laser does not hit a blood vessel, the Soldier may see only flashes of light, followed quickly by some painless loss of vision. If the laser damage areas of the eye responsible for peripheral vision, the Soldier may never recognize a visual deficit. If the Soldier was looking directly at the laser source however, there will likely be a major loss of visual clarity. With simple retinal burns in the retina's periphery, most of the visual symptoms recover with hours to days of rest, reassurance, and nonspecific treatment the same as with COSR. Calm, professional treatment at each level of medical care should emphasize that the injury is not vision-threatening and the chances for some, if not total, recovery is good. Soldiers with the simple retinal burns should provide self-care to decrease the risk of assuming a patient mind-set and to promote their chances of RTD. For additional information on the management of laser eye injuries, refer to FM 8-50.

Middle Ear Injuries/Diseases

8-34. Temporary loss of hearing can be cause by a decreased acoustic sensitivity following a brief extremely intense noise (explosive) or less intense, longer duration noise. Tinnitus (ringing in the ears) can also result from acoustic nerve damage or irritation, as well as from high doses of certain drugs, such as aspirin. Distinguishing physiologic from psychogenic hearing loss may require consultation with an otolaryngologist (ears, nose, and throat specialist).

Peripheral Neuropathies

8-35. Peripheral neuropathies include compression neuropathies, which are especially likely in military settings (for example, *rucksack palsy*). Depending on severity, they may require temporary job reclassification during convalescence. As these injuries are not life-threatening, a hasty diagnosis should not precede a trial of Soldier restoration treatment.

Uncommon Endemic Neurologic Disorders

8-36. These physical diseases can first manifest with cognitive emotional and/or behavioral symptoms. A comprehensive neurological examination is required for the definitive diagnoses. Examples include—

- Guillain-Barre Syndrome manifests with muscle paralysis, usually without sensory loss, which ascends the legs and arms, then the trunk, over hours to days. It is sometimes triggered by immunizations, as might be given to troops deploying overseas. It often progresses to a life-threatening situation as the muscles of respiration become involved. This requires evacuation to EAC and continental United States (CONUS). Fortunately, recovery is usually complete, but it takes months to years.

- Multiple sclerosis is a disease that can mimic many types of COSR/BH disorders with its sometimes transitory, shifting motor, sensory, speech, and cognitive/emotional symptoms. It is made worse by stress and may be difficult to diagnose. Once confirmed, multiple sclerosis cases should be evacuated to CONUS, as should other rare, progressive diseases like Lou Gehrig's disease (amyotrophic lateral sclerosis).
- True convulsive seizure can manifest after head injury or a sublethal or chronic nerve agent exposure. These are treated with normal anticonvulsant medications. Fear of nerve agent exposure may lead some Soldiers to experience psychogenic seizures. These psychogenic seizures are also called pseudo-seizures. In addition to falling unconscious and convulsing, urinary and fecal incontinence can occur during a pseudo-seizure.

SUBSTANCE ABUSE/DEPENDENCE

ALCOHOL

8-37. Substance abuse is an example of misconduct stress behaviors and not necessarily COSR. Drug and alcohol abuse may occur in active CZs and nearby areas where use is explicitly prohibited and severely punished. Personnel performing the COSC assessment should be familiar with evaluation and treatment of substances abuse and dependency. Combat operational stress control providers should consider the following:

- Heavy habitual use of alcohol, even by otherwise capable officers and NCOs, may go unnoticed in peacetime. However, alcohol abuse it may degrade necessary mission performance demanded by combat and may result in withdrawal symptoms when access to alcohol is interrupted. Alcohol withdrawal is potential medical emergency; consultation with medical personnel is essential.
- Intoxication or withdrawal from alcohol, barbiturates, and tranquilizers may be mistaken for COSR or another BH disorder. Intoxication or withdrawal requires medical treatment. Withdrawal seizures or impending or ongoing delirium tremens need emergency medical treatment.

OVERUSE OF STIMULANTS

8-38. Stimulants may cause panic attacks, hyperactivity, mania, rage attacks, psychosis or paranoia. Cessation of amphetamines after prolonged use causes a crash characterized by extreme sleepiness, lethargy, overeating depression and suicidal thinking. This condition may require one to two weeks of hospitalization to assure safe recovery.

HALLUCINOGENIC DRUGS

8-39. Hallucinogenic drugs cause sensory distortion, panic, bizarre thoughts, and potentially dangerous behaviors. These drugs may be employed by the enemy as chemical or biological warfare agents. Phencyclidine hydrochloride (PCP) is especially problematic since it also blocks pain and tends to make those under its influence paranoid, violent, and abnormally strong. Hallucinogenic drug psychosis should not be treated with antipsychotic drugs. Physically restrain and sedate patients as necessary.

INHALATION OF FUMES

8-40. Inhalation of fumes (either by accident or as deliberate abuse) and carbon monoxide poisoning can cause disoriented, abnormal behavior. Supportive treatment and, specific antidotes/medication may be needed.

ANTICHOLINERGIC DELIRIUM

8-41. In combat, atropine delirium may occur. Soldiers are equipped with atropine injectors to use as first aid against nerve agents. Two milligrams (mg) (one atropine injector) without nerve agent challenge can cause rapid pulse, dry mouth, slightly dilated pupils, decreased sweating (hot, dry, flushed skin), and urinary retention. In some individuals, 6 mg of atropine (equal to three atropine injectors) may cause hallucination and disorientation in the absence of a nerve agent challenge. Such side effects may be more pronounced in sleep-deprived Soldiers. Overheated Soldiers are more susceptible to the atropine side effects. Atropine compounds the complications of overheating by diminishing the body's ability to lose heat through sweating. One dose (2 mg) of atropine can reduce the efficiency of heat-stressed Soldiers. Two doses (4 mg) will sharply reduce combat efficiency, and 6 mg will incapacitate troops for several hours. Some plants can also cause anticholinergic delirium when eaten.

ANTICHOLINESTERASES

8-42. A nerve agent is an anticholinesterase similar to many insecticides. Low-dose nerve agent exposure may produce miosis (pinpoint pupils) without other signs. Miosis decreases vision except in very bright light and may cause eye pain when attempting to focus. This miosis may take hours to days to improve spontaneously, depending on the degree and type of exposure. Evidence gathered from farm workers poisoned by insecticides suggests that mild personality changes, insomnia, nightmares and chronic persistent depressive symptoms may be seen even after use of an antidote. Low-dose nerve agent exposure may lower the seizure threshold of many Soldiers. True epileptic seizure cases must be distinguished from those Soldiers who may have pseudo-seizures.

BEHAVIORAL DISORDERED PATIENTS IN THE THEATER

PRIMARY BEHAVIORAL DISORDERS

8-43. Primary BH disorders (especially schizophrenic-form/schizophrenic disorder, major depression, and bipolar disorder) will continue to occur at approximately the same rate as in peacetime. Some Soldiers may hide their disorders by receiving care through civilian channels. Once in the theater they may experience a relapse or self-refer to a MTF when their medication supply is exhausted.

PERSONALITY DISORDERS

8-44. Preexisting personality disorders may make a Soldier unable to adapt to military life. However, studies have failed to show a relationship between personality disorders and the likelihood of breakdown in combat. Once Soldiers with personality disorders have developed COSR or a BH disorder, they may have greater difficulty recovering and RTD. Diagnosis should never be made in haste; diagnostic criteria must be supported with adequate historical information.

Chapter 9

Combat and Operational Stress Control Stabilization

SECTION I — EMERGENCY STABILIZATION

STABILIZATION

9-1. Emergency stabilization is the acute management of disruptive behavior resulting from COSR and/or a behavioral disorder. The disruptive behavior severely impacts unit functioning by posing a danger to self and/or others. In some cases, an underlying medical condition leads to the disruptive behavior and may present an additional threat to the Soldier's life. Emergency stabilization consists of interventions that temporarily reduce a disturbed Soldier's threat of self or others, thereby allowing further medical evaluation and/or treatment. Some behavioral disorders are associated with violent behavior, such as psychotic disorders, bipolar manic disorders, antisocial personality disorder, and borderline personality disorder. Violent behavior is also associated with disruption of brain functioning due to organic factors such as intoxication, hyperthermia, metabolic imbalance, or CBRN exposure.

9-2. The COSC triage process will be repeated throughout the emergency stabilization and will determine the disposition of the Soldier. After emergency stabilization, subsequent triage could result in an immediate RTD, transfer to a COSC Soldier restoration program for observation, or further evacuation.

METHODS USED FOR EMERGENCY STABILIZATION

9-3. Methods that may be used for emergency stabilization include—

- Providing verbal reassurance and reorientation are the best methods for controlling an agitated or disruptive Soldier. If these fail, a nonthreatening show of strength or force may suffice, or sedating medications may be offered to the Soldiers. If all other means fail to reduce the threat to self and/or others, physical restraint must be considered. Given the risk for violence, it is inadvisable to attempt subdue/restraint method one-on-one.
- Applying physical restraints is reserved for subduing and restraining agitated or disruptive Soldiers who fail to respond to safer and less restrictive forms of restraint (for example, verbal warnings or show of strength). Placing a disturbed Soldier into physical restraint increases the risk of injury to the Soldier and restraint team. Prolonged or improper application of physical restraint can cause injury to the disturbed Soldier. Given the potential for injury, it is paramount that COSC personnel receive training in proper physical restraining methods. Safe medical evacuation by ground (preferred) or air ambulance.
- Regardless of the method, the restrained Soldier must be checked frequently to guard against nerve injuries or impaired circulation, which may lead to skin ulcers or gangrene. It is important to check periodically to ensure the Soldier is not secretly escaping from restraints. The Soldier is provided verbal reassurances with positive expectations for his recovery each time he is checked.
- Chemical restraints (such as medication) can be administered to a disturbed Soldier to reduce the risk of harm to self or others. Medication may be offered to the Soldier in conjunction with verbal reassurances and reorientation. Chemical restraints may only be ordered by a medical professional who is authorized to prescribe medication when a Soldier is incompetent to make medical decisions for himself and/or when the Soldier's behavior places himself or others in danger. Once administered, medical personnel must observe for side effects and adverse reactions, and must consider administering additional medications as needed. When a Soldier is in physical restraints, medication may no longer be essential, but may serve to reduce the risk of escape, limb damage, and overheating. As a secondary benefit, once the medications reduce the

Soldier's agitation, others in the vicinity may feel safer and calmer. Before prescribing an antipsychotic medication, there are a few things to consider. First, some antipsychotic drugs may take several hours or days to take effect. Second, early administration of an antipsychotic drug may confuse the clinical picture for the next evaluator in the evacuation chain. The recommendation for most cases is to use no medication unless it is truly necessary for management.

SECTION II — COMBAT AND OPERATIONAL STRESS CONTROL FULL STABILIZATION

FULL STABILIZATION

9-4. Full stabilization is normally the mission of the Level III MTF such as a CSH specialty clinic's psychiatric service. Full stabilization goes beyond securing the safety of the Soldier and those around him. It provides a safe environment for the Soldier to receive treatment interventions, continued evaluation, and assessment for RTD potential. If RTD within the evacuation policy is not feasible, the full stabilization process helps to prepare the patient for a safe, long-distance evacuation. If a Level III MTF is not available, full stabilization may be accomplished by COSC personnel when appropriately supported or by using equipment diverted from the COSC Soldier restoration capability. Considerations for full stabilization may include—

- Conducting full stabilization for NP patients is desirable for the sake of the Soldier's future treatment and for the potential of returning some Soldiers to duty. However, full stabilization is personnel intensive with a relatively low RTD payoff. Providing only sufficient stabilization to allow evacuation from the theater may be acceptable in order to maintain the other COSC functions.
- Ensuring appropriate timely evacuation of Soldiers with NP/behavioral disorders according to the theater evacuation policy. It is preferred that full stabilization is achieved for all NP patients to facilitate appropriate and timely evacuation according to the theater evacuation policy.
- Assessing and triaging of COSR Soldiers undergoing full stabilization is an ongoing process. In subsequent triages, if a Soldier becomes stable the potential for RTD, they may be transferred to a COSC Soldier restoration or reconditioning program, or may be RTD directly.

TENETS OF FULL STABILIZATION

9-5. The COSC full stabilization includes ongoing evaluation of RTD potential. This requires assessment of mental status and performance capability overtime without excessive drug effects or limitation on activity. Contact with the Soldier's unit may be needed to get information on prior history and functioning. The further from the unit the Soldier has been evacuated, the more difficult it may be to contact the Soldier's unit. Full COSC stabilization normally takes several days. To the extent compatible with safety, the stabilization program should adhere to the principles and methods for treating COSR and behavioral disorder (such as BICEPS and 5 R's).

9-6. During full stabilization, special efforts should be made to maintain and reinforce the Soldier's identity as a Soldier. Techniques that may be helpful in maintaining the Soldier mind-set include—

- Keep Soldiers in duty uniform, not pajamas, as soon as they can be allowed.
- Maintain rank distinctions and appropriate military courtesy.
- Encourage self-care and helping behaviors.
- Engage in military work activities appropriate to the Soldiers' level of function and MOS.

9-7. Initial and ongoing assessments are essential to tailor the treatment to the Soldier's individual needs. It is essential that clinical documentation is available for these assessments. The Soldier's condition is an evolving one, and must be monitored throughout full stabilization. If assessed capable to RTD, efforts should be made to return the Soldier to duty.

9-8. Ongoing treatment and/or therapeutic modalities are essential to improving a Soldier's chances to RTD whether in theater or after evacuation. Therapeutic modalities are similar to those used on inpatient units, but must remain consistent with COSC principles. These modalities include medication, individual psychotherapy, group psychotherapy, and appropriate therapeutic occupations. Observed responses to therapeutic modalities allow informed recommendation for RTD status.

FULL STABILIZATION FACILITIES

9-9. Full stabilization is commonly conducted in the CSH. The CSH can provide more sophisticated procedures, laboratory and x-ray capabilities than are available at a medical company MTF. If a medical unit, CSC is providing emergency or full stabilization, the Soldier must be kept separate from other Soldiers in Soldier restoration or reconditioning.

9-10. If the Level III MTF (CSH) cannot provide sufficient inpatient psychiatric treatment for Soldiers requiring stabilization and preparation for evacuations, the theater/AO COSC consultant may recommend up to two temporary COA until the shortfalls resolve. He can recommend to the higher medical command that COSC personnel from one or more medical unit, CSC augment the Level III MTF psychiatric service until the caseload decrease or until replacements or additional COSC personnel are brought into theater or the MTF. Lastly, the medical C2 headquarters may direct that a COSC Soldier restoration or reconditioning assets be collocated with the Level III MTF to provide an "overflow" ward as well as augmenting staff. Definitive information is provided on the CSH NP ward staff capabilities in FM 4-02.10.

9-11. The MF2K CSH has a psychiatrist, three psychiatric nurses and a medical/surgical nurse, a social worker, nine mental health specialists and an OT sergeant at full strength. It has a modular hospital tent to provide a 20-bed hospital ward. The MF2K hospitals were designed to be at one location, but may follow the MRI doctrine of splitting off a portion to a remote site, taking with it some of the psychiatry service personnel. One portion would not have the psychiatry ward module, and would need to function similarly to the MRI CSH psychiatry service.

9-12. The MRI CSH has a psychiatrist, a psychiatric nurse, and two MH specialists. The service does not have an organic psychiatric ward where NP patients are admitted. There is normally an intermediate care ward with medical ward staff designated to receive NP patients. Some of these patients seen by the NP clinic will already be stable, and are being evaluated and prepared for RTD or evacuation. Most patients who are admitted for emergency or full stabilization will be on the designated ward, although a few with serious medical complications could require the intensive care ward. If the hospital has a high census of medical and surgical patients, most of the NP patients will need to occupy beds on the intermediate care ward. The MRI hospital may be augmented by staff from a medical company/detachment, CSC to operate a NP ward, as required.

9-13. Full stabilization facilities in theater can be categorized into two types:

- Mobile facilities use general purpose (GP) large or medium tents as used in medical units, CSC and can also be available to a CSH or a tent, expandable, modular, personnel (TEMPER) tent as used in a CSH.
- Fixed facilities use buildings that were previously hospitals or buildings converted to hospitals.

9-14. Facilities used for full stabilization include tents, Deployable Medical Systems (DEPMEDS) TEMPER tents, and fixed facilities. The adaptation of these facilities have both advantages and disadvantages that include—

- The principal advantage of the (hospital) TEMPER tents, as assembled into DEPMEDS hospital, is their climate control capability. This may be a significant safety advantage for treating seriously disturbed patients in restraints with high-dose medication, which can disrupt body's ability to regulate normal body temperature. The TEMPER and standard tents both pose greater problems for security than do fixed facilities. The staff may, therefore, have to rely more than is ideal on physical restraints and medications for sedation of some cases. Blankets or screens can be used to isolate or segregate problem patients from others. Such partitions reduce mental contagion but provide little true protection.

- Standard hospital beds are on high, lightweight metal legs with wheels. For full stabilization purposes, these should be replaced with standard low, stable cots to hold strong, agitated patients in restraints. The cots also make a more "military" setting and can be used as seats for group activities.
- When feasible, it is best to have a separate "closed" (high security) and "open" (moderate/minimal) security area. The latter could be a standard GP large tent (the same as those of the minimal care wards [MCWs] located close to the TEMPER tent of the official NP ward). The specialists (MOS 68WM6 [practical nurse] and 68W [health care specialist]) of the MCW could be given on-the-job training in supervision and military group activities for the moderate/minimum security cases if the NP staff is too small.

Chapter 10

Combat and Operational Stress Control Soldier Restoration

SECTION I — SOLDIER RESTORATION

SOLDIER RESTORATION PROGRAM

10-1. Soldier restoration is normally a 24- to 72-hour (1- to 3-day) program in which Soldiers with COSR receive treatment. Soldier restoration is accomplished using the principles of BICEPS and the 5 R's as discussed in the previous chapter. The 5 R's are tailored to the needs of the Soldier. Soldier restoration is conducted as close to the Soldier's unit as possible. Soldier restoration can be conducted by medical units/elements throughout the theater with the assistance of organic and/or augmented COSC personnel. The medical company/detachment, CSC is staffed and equipped to establish Soldier restoration programs. Soldier restoration may also be accomplished at a Level III MTF (CSH). The tenets remain the same but implementation may differ in the duration of Soldier restoration and the specialized skill and knowledge of available providers. There are three subcategories of Soldier restoration (referred to as lines of Soldier restoration) based on location and available resources that are discussed in Section II of this chapter.

SOLDIER RESTORATION PROCESS

10-2. The process of Soldier restoration involves several steps that include screening, assessment, and interventions, reintegration/coordination, and movement of Soldiers. A discussion of each of the steps in the process is provided below.

SCREENING

10-3. Adequate medical screening and treatment as necessary must be done before the Soldier begins the Soldier restoration program. Minor medical conditions can be treated during routine sick call. Soldiers entering a Soldier restoration program should be only those *hold* cases that require continuous medical and/or BH evaluation and observation for 24 hours or more. The criteria for *hold* cases are discussed under COSC triage in Chapter 8. Tending to and restoring physiological status (such as sleep and hydration) is a priority. Some Soldiers who need Soldier restoration will also have a mental disorder. Treatment for the mental disorder may continue or be initiated during the Soldier restoration process.

ASSESSMENT AND INTERVENTION

10-4. Initial assessment and subsequent COSC interventions depends upon the severity of the COSR and to what extent the reactions interfere with the Soldier's ability to function. More thorough interviews are conducted only after the Soldier's physiological status has been restored.

REINTERGRATION

10-5. Coordination is required to assist Soldiers with reintegration back into their units. Soldiers are reassessed regularly. When a Soldier experiencing COSR begins to improve, the COSC interventions shift toward the reintegration of the Soldier back into his unit. The COSC personnel must work with the COSC consultant of their area or other resources to assist with this reintegration.

MOVEMENT

10-6. Movement of Soldiers from one line of Soldier restoration to another that has greater capabilities or security may be required. When movement of these Soldiers is required, vehicles used to transport them should be nonambulance vehicles, if possible, and accompanied by escorts from their unit. If nonmedical unit members are not available to perform escort duties, medical augmentation may be required.

PRINCIPLES AND PROCEDURES OF SOLDIER RESTORATION

10-7. Initial Soldier restoration begins as close to the Soldier's unit as possible, normally near an MTF where the Soldier can get a reprieve from extreme stress but at the same time can be close to his unit. Normally, Soldier restoration facilities are established in the BSA near or adjacent to the FSMC/BSMC Level II MTF. Soldier restoration is not feasible at locations that are consistently under artillery, air, or direct-fire attack, but ideally are still within the sound of the artillery or other reminders of battle. If there is potential for attack, there must be cover and defensive positions. The location should not be one from which a move is likely within 24 hours. If there is a significant possibility of a move, only those cases that can participate actively in the move with minimal supervision should be managed at this location. The specific site of the Soldier restoration facility should be adjacent to the medical company's area and be out of the immediate (close) sight of the medical triage and treatment areas

REASSURANCE

10-8. Immediate reassurance is given to the Soldier with COSR beginning with the COSC triage as discussed in Chapter 8. Tell the Soldiers that they are temporarily joining the unit, not as patients, but as Soldiers who need a few days to recover from COSR. Emphasize that COSR is a normal response to extremely abnormal conditions. Rapid recovery is also normal and RTD is expected. Reassure the Soldiers about safety and what to do in the event of an attack or march order. Lastly, orient the Soldiers to the Soldier restoration program.

STRUCTURED MILITARY ENVIRONMENT

10-9. An emphasis should be placed on maintaining a highly structured military unit environment and schedule of activities in order to keep the Soldier from adopting a patient role. Assign the Soldier to a squad under supervision of a specific squad leader. The squad leader may be a CSC unit/MH section member or a member of the BSB subordinate unit. The squad leader may also be one of the Soldiers there for Soldier restoration, if their condition allows. The newly arrived Soldier is assigned to a tent that has been designated/erected for the Soldier restoration program. Soldiers with more dramatic COSR or BH symptoms should be temporarily quartered separately from other Soldiers receiving Soldier restoration. In remaining consistent with the principle of treating Soldiers with COSR as Soldiers instead of patients, it is recommended that Soldier restoration facilities do not display the Red Cross as displayed on the MTF. This could affect their protection under the Geneva Conventions. See FM 4-02 for information on the Geneva Conventions.

SUPPORT THE SOLDIER'S MILITARY IDENTITY

10-10. Maintain appropriate rank distinctions, titles, and military courtesies within the confines of the tactical situation. Expect the Soldier to maintain military bearing and personal appearance. They should be in duty uniform. Conduct basic Soldiering skills. Do not take personal possessions away from the Soldier. This includes weapons, unless there is significant concern for the Soldier's safety. If the Soldier arrives with a weapon, SOPs of the AO will guide whether or not the weapon is secure for the Soldier. Also, encourage the unit to maintain contact with its Soldier.

REPLENISHMENT OF PHYSIOLOGICAL STATUS

10-11. Get the Soldier under shelter and cool down if overheated, warm up if cold, and dry off if wet. Providing hot beverages and/or soup will also assist with restoring body temperature. Replenish hydration with palatable beverages and meals. Unless the Soldier is totally exhausted, institute some personal hygiene. If the Soldier needs uniforms or equipment, coordinate with the BSB S4 or supporting logistic element. Restorative sleep should be as normal as possible. Soldiers are typically able to fall asleep when reassured of safety, and do not routinely require medication to induce sleep. They should be informed that they may awaken with vivid and frightening dreams, and be instructed on quick relaxation techniques to go back to sleep. If those measures are insufficient, one-time medication with a sedative/hypnotic may be considered, coordinated, and prescribed by a physician or PA from the supporting medical company. The duration of sleep should be sufficient to make substantial progress in repaying the sleep debt. It should also begin the process of restoring a reasonable sleep/wake cycle that is consistent with the Soldier's duties in his unit.

RESTORE CONFIDENCE

10-12. Restoring a Soldier's confidence may include—

- Providing regularly scheduled formations to keep Soldiers informed of daily activities and the tactical situation, including information about their unit, when available.
- Providing therapeutic occupations that are based on the Soldier's current functional ability. The OT (or COSC team members under the guidance of the OT) selects therapeutic occupations that support the Soldier's military identity and enhance the Soldier's sense of competence. Therapeutic occupations may include activities of daily living (such as physical fitness or uniform maintenance); educational activities (selected common and collective Soldier task training or life skills training); work/productive activities (militarily relevant tasks such as vehicle maintenance or site maintenance); leisure/recreational activities (enjoyable, relaxing activities, games, and sports); social participatory activities (cooperative/competitive sports, games, ceremonies, or celebrations). Therapeutic occupations provide the casualty with a challenge but afford successful performance that shows the Soldier that he is still capable and competent. This realization plays the dominant role in restoring the Soldier's sense of confidence, functional capacity, and ability to RTD.
- Training and teaching Soldier on methods for managing excessive stress.
- Facilitating factional review of precipitating event that has caused the Soldier's COSR through talking (ventilating) and coaching. Combat and operational stress control personnel (or health care providers trained by COSC personnel) facilitate these discussions. They help Soldier restore perspective with questions and coaching. The factual review (debriefing) is often best done individually unless several Soldiers experienced the same event.

SECTION II — LINES OF SOLDIER RESTORATION

FIRST-LINE SOLDIER RESTORATION

10-13. First-line Soldier restoration is usually provided at the BSMC/FSMC in the BSA. It is provided by personnel organic to the BSMC/FSMC usually assisted and supervised by COSC personnel organic to the medical company and/or from CSC unit personnel. This also applies to a cavalry regiment. Because of the high mobility of the BSMC/FSMC, Soldier restoration will often be a 24- to 72-hour process. First-line Soldier restoration also may occur in relative secure locations in division, corps, and theater AOs. First-line Soldier restoration in the division is provided near and supported by a Level II MTF such as a MSMC with its organic MH and/or minimal augmentation from a CSC unit. First-line Soldier restoration in corps and theater is provided by ASMCs with responsibility for their AO. It is provided by COSC personnel organic to the medical company and/or from CSC unit personnel. The MSMC and ASMCs will be relatively unlikely to move on short notice so Soldier restoration up to 72 hours should be feasible. Support provided for Soldier restoration is a responsibility of the MH sections of supporting medical companies. Tasking for support of Soldier restoration programs is accomplished through the command surgeon.

SECOND-LINE SOLDIER RESTORATION (SOLDIER RESTORATION CENTER)

10-14. Each of the first-line Soldier restoration locations listed above should be backed up by a second-line Soldier restoration capability at a location that is less likely to have to move on short notice and has more COSC capabilities. Soldier restoration centers may be located near the MSMC, ASMC, or with a CSC medical company or detachment. They can be located in the BSA near the BSMC/FSMC when it is at a stable base camp. The Soldier restoration center may receive *hold* cases that are transferred from the first-line medical companies, as well as Soldiers from nearby units. Second-line Soldier restoration has more equipment and a greater range of COSC expertise. This permits a 72-hour holding capacity for a stable well-organized Soldier restoration center and may provide full stabilization.

THIRD-LINE SOLDIER RESTORATION

10-15. In some scenarios, units with Soldiers in need of Soldier restoration may be significantly closer to a Level III MTF such as a CSH than to a medical company or CSC unit. On order, a Soldier restoration program may be conducted by the CSH specialty clinic NP staff, which may also be augmented with personnel from a medical company/medical detachment, CSC. In such cases, a Soldier restoration program may be conducted by the CSH specialty clinic NP staff. It may be augmented with personnel from the medical company or detachment, CSC. Soldier restoration near a CSH must be kept clearly separate from the patient ward. It may be done at a MCW or in separate tents.

RETURN TO DUTY OF RECOVERED COMBAT AND OPERATIONAL STRESS REACTION SOLDIERS

10-16. Most COSR symptoms do not necessarily improve completely while the prospect of combat continues. The Soldiers should be given the positive expectation that they will RTD. Every possible effort should be made to return Soldiers to their original unit. The RTD of Soldiers following Soldier restoration depends on how near the Soldier's unit is, the availably of a means of transportation, and the tactical situation. Ideally, their units are contacted to send someone to get the Soldier or he may be returned to his unit by way of the personnel replacement company. Mental health personnel coordinate directly with the unit to which the Soldier is returning and/or with COSC, other medical personnel, or UMT supporting the unit's AO. These contacts can consult with the leaders of the Soldier's unit and facilitate the Soldier's acceptance back into his unit.

DOCUMENTATION

10-17. Soldiers receiving Soldier restoration interventions must be tracked from the initial contact until they are returned to their unit. A record must be maintained of interventions and activities provided during Soldier restoration, as well as the Soldier's response. A notation of the dates, any pertinent medical data, and providing unit is entered in the Soldier's medical record. An administrative summary of Soldier restoration services may be developed, stored, and disposed of as directed by the Soldier restoration center SOP. If the Soldier received BH treatment, documentation should be maintained as delineated in AR 40-66. A statistical record is maintained and sent to higher medical headquarters according to SOP.

Chapter 11

Behavioral Health Treatment

SECTION I — BEHAVIORAL HEALTH CARE

TREATMENT FOR BEHAVIORAL DISORDERS

11-1. Behavioral health treatment exists when there is an explicit therapist-patient or therapist-client relationship. Behavioral health treatment is provided for Soldiers with behavioral disorders to sustain them on duty or to stabilize them for referral/transfer. This is usually brief, time-limited treatment as dictated by the operational situation. Behavioral health treatment includes counseling, psychotherapy, behavior therapy, occupational therapy, and medication therapy. Treatment assumes an ongoing process of evaluation, and may include assessment modalities such as psychometric testing, neuropsychological testing, laboratory and radiological examination, and COSC providers' discipline-specific evaluations.

COMBAT AND OPERATIONAL STRESS CONTROL BEHAVIORAL HEALTH PROVIDER

11-2. There are five professional disciplines and two enlisted specialties that serve as BH/COSC providers, see Chapter 3. Army BH/COSC providers are officers who hold active individual clinical privileges granted by an MTF according to AR 40-68. Providers also include enlisted medical specialist who do not hold individual clinical privileges, but who work under direct supervision of a professional COSC provider. Behavioral health treatment is provided within the scope of the providers' clinical privilege, training, and experience.

SECTION II — BEHAVIORAL HEALTH TREATMENT PROTOCOLS AND MEDICATIONS

BEHAVIORAL HEALTH TREATMENT FOR SOLDIERS

11-3. Behavioral health treatment is provided to Soldiers with diagnosed behavioral disorders (see Chapter 8), and who require more intentions for their diagnoses. It is both inappropriate and detrimental to treat Soldiers with COSR as if they are a BDP. A therapeutic relationship may promote dependency and foster the "patient" role. Likewise, medication therapy and the highly structured treatment modalities imply the "patient" role. Medication for transient symptom relief (insomnia or extreme anxiety) may not be detrimental if there is no expectation that medication will continue to be prescribed.

BEHAVIORAL HEALTH PROTOCOLS AND PROCEDURES FOR TREATMENT

11-4. Behavioral health treatment is provided within a context of preventing symptoms of behavioral disorder from progressing in severity and improving the level of functioning. The principles of BICEPS apply. Treatment should not impede activities for other prevention levels (for example, universal, selective, and indicated). Treatment should not interfere with the Soldier's duties or the unit's mission. The COSC provider should emphasize Soldier's identity as a Soldier despite having a diagnosis. Transportation of the Soldier to the provider (or vice versa) and ability to assure reliable meeting times can be limiting factors, depending on the operational environment. Medication refills must be coordinated for availability.

Standards of Treatment

11-5. Treatment standards are the same in the deployed environment as in garrison. When operational requirements dictate that clinical standards of treatment/care are waived or relaxed, it must be approved by the AO COSC consultant. (AR 40-68 provides additional information.) Treatment should be tailored to the anticipated availability of the Soldier and COSC provider. Short-term interventions are more practical than long-term commitments. If longer-term treatment is necessary, design the intervention in time-limited modules. Under no circumstances should treatment diminish the Soldier's ability to provide self-care and to defend himself. Exceptions include emergency stabilization and preparation for evacuation. In addition, the VHA/DOD Clinical Practice Guidelines website (http://www.oqp.med.va.gov/cpg/cpg.htm) offer clinicians evidence-based assessment and treatment algorithms for acute stress disorder, PTSD, and many other behavioral disorders.

Army Regulations Governing Evaluations

11-6. Fitness for duty evaluations are conducted as necessary within the priorities of the supported commanders according to AR 40-501; psychiatrists should not initiate a medical evaluation board without first ensuring the Soldier has received adequate treatment. This treatment may not be available in theater. Command-directed evaluations are conducted as necessary within the priorities of the supported commanders according to DODD 6490.1 and MEDCOM Regulation 40-38. Clinical documentation should be safeguarded according to AR 40-66 and local command policy. Treatment should be conducted in a location that is as private as possible. Information can be released to a third party if the Soldier consents. Combat and operational stress control providers need to notify command when the Soldier's safety is in question (suicidal, homicidal) or if the Soldier is removed from his unit for medical observation. Additional release of information to command is on a need-to-know basis. For any questions on release of information on COSR and NP patients, consultation with supporting JAG office is advised. Also, maintaining the Soldiers health record with clinical data is required and accomplished according to AR 40-66 and supplemental theater policy as appropriate. Treatment should occur throughout the evacuation process and follow-up is expected at home station.

Chapter 12

Reconditioning

SECTION I — RECONDITIONING PROGRAM

LOCATION

12-1. Reconditioning programs are intensive efforts to restore those Soldiers triaged as a *refer* case, but who still have good potential for RTD. Referral to reconditioning can be from Level II MTF, COSC Soldier restoration program, or from Level III MTF. Reconditioning includes the rehabilitation of Soldiers with mental disorders, such as substance abuse/dependency. Reconditioning programs are conducted by COSC personnel and Soldiers usually participate for up to 7 days. Soldier participation may be extended by a case-by-case exception to theater evacuation policy. Reconditioning is conducted only in a theater where there are adequate COSC elements and supplies. When COSC resources are needed for other functional areas, any ongoing reconditioning program reduces its scope of services or closes. Reconditioning is provided at first-, second-, or third-line centers determined by location and available resources.

PROVIDERS

12-2. Conducting reconditioning programs is a mission of the medical company, CSC and the medical detachment, CSC (MRI). Reconditioning can also be done on a small-scale by a CSC medical detachment (MF2K) under some circumstances.

SECTION II — TENETS AND PROCEDURES OF RECONDITIONING

RECONDITIONING

12-3. Reconditioning may be considered an extension of Soldier restoration. Reconditioning is similar to Soldier restoration, but with potentially longer stay, treatment strategies focus on preventing atrophy of skills and assisting Soldiers in regaining skills needed for duty. Reconditioning makes more use of BH treatment modalities.

STRUCTURED MILITARY ENVIRONMENT TO SUSTAIN SOLDIER'S IDENTITY AS A SOLDIER

12-4. Like Soldier restoration, reconditioning emphasizes a highly structured military unit environment and schedule of activities in order to keep the Soldier from adopting a "patient" role. Maintaining a military environment is even more critical when colocated with hospitals or other service support elements.

REPLENISHMENT OF PHYSIOLOGIC STATUS AND CONFIDENCE

12-5. Reconditioning initially emphasizes physical replenishment and hygiene, but later shifts the emphasis to more closely match the conditions that the Soldier should expect when RTD. In order to restore confidence in the Soldier, unit formations are held on a regularly scheduled basis, and include the Soldiers receiving reconditioning.

OCCUPATIONAL THERAPY

12-6. Therapeutic occupations may include—

- Activities of daily living (for example, hygiene, physical fitness, and uniform maintenance).

- Educational activities (for example, common and collective Soldier task training, and life skills training).
- Militarily relevant work details according to the current functional level and MOS/duties of the Soldier.
- Leisure/recreational activities (for example, enjoyable, relaxing activities, including cooperative physical and mental activities, and basic relaxation techniques).
- Social participatory activities (for example, games and ceremonies).
- Group training in relaxation techniques.
- Ventilation and coaching where COSC (or medics trained by COSC personnel) encourage discussion about stressors and their impact on mental well-being. This technique helps the Soldier to restore personal perspective with questions and coaching. Individual counseling and therapy may improve the Soldier's functions as well.

12-7. When reclassification recommendations are considered, an adjacent CS or CSS unit may be able to provide a job that will match the Soldier's abilities. This provides an opportunity to demonstrate the Soldier's abilities and build confidence. The reconditioning personnel may recommend to S1/G1 that the Soldier be reclassified to another MOS.

FIRST-LINE RECONDITIONING PROGRAM

12-8. First-line reconditioning programs in the corps are staffed by task-organized CSC elements from the CSC company or by an MRI CSC detachment. Reconditioning can be conducted by an MF2K CSC detachment, but would preempt its Soldier restoration capability. If the inpatient NP workload is light, a small reconditioning center may be staffed by personnel from the NP ward and consultation service of an MF2K CSH. The first-line reconditioning center usually collocates with a CSH but must maintain its separate, nonhospital identity. It should not be situated among the hospital wards, or near the morgue, triage area, or helicopter pad. Within a theater/AO, the preferred option is to have one or two reconditioning centers placed such as to allow easy access from Soldier restoration centers. Under major, prolonged combat conditions, it may be more appropriate to have one reconditioning center behind each division. The reconditioning center is dependent on the hospital for support. The reconditioning center works with the supporting hospital by sending work parties of Soldiers in the program to assist in food preparation and delivery and cleanup chores. These Soldiers may also be used for assisting with work details throughout the hospital, but must be under direct supervision of either hospital or BH personnel. The reconditioning center uses the supporting hospital's medical records section to maintain the permanent case records. Cases in the reconditioning center are counted as patients in the reconditioning center on the daily hospital census. The cases are not counted as occupied beds when reporting the hospital bed occupancy. Upon disposition from the reconditioning center, whether for RTD, retraining for other duty, or evacuation, the reconditioning center prepares the chart for further evacuation or writes the discharge summary and closes the hospital's chart.

12-9. Some reconditioning cases will be able to return to far forward CS or CSS duty. However, many of the Soldiers who need reconditioning will be unable to return to their original unit, due to combat operations or the nature of their symptoms. Soldiers who undergo a 4- to 7-day Soldier restoration program in the corps should not be crossed off the division's personnel rolls (as specified in AR 40-216). Continuing contact with the unit increases RTD rates. For Soldiers who recover but are no longer on the division's rolls, every feasible effort should be made to return them to their original units. The chief, division MH and the CSC unit teams that are attached to reinforce the division should maintain frequent contact with the reconditioning centers that support the division. Coordination efforts, through the division support command (DISCOM) sustainment brigade and the personnel replacement system by division BH personnel, should attempt to facilitate return of recovered Soldiers to their original units. If recovered Soldiers cannot be returned to their previous small unit, consider forming them into cohesive pairs or small groups which can be reassigned to a new unit together. The standard corps evacuation policy is seven days, but some Soldiers with good potential for RTD may need a few more days at the reconditioning center. The corps evacuation policy is at the discretion of the theater commander. The principle purpose of a short evacuation policy is to keep hospital beds available for mass casualties and to minimize the expense and labor required to treat serious surgical and medical cases.

SECTION III — EVACUATION POLICY

EXTENDING THE THEATER EVACUATION POLICY

12-10. If need assessment indicates that a longer period of time (more than seven days) will achieve a Soldiers' RTD, the theater COSC consultant can request permission to extend the theater evacuation policy (up to 14 days). He submits the request for authority via the commander, MEDBDE, through the corps surgeon to the corps commander. The following facts apply: The reconditioning program, because of its austerity, is not a significant logistical burden to the corps. Soldiers in reconditioning perform useful work details and perimeter defense. Lastly, the increase in RTD is important in operations such as stability and reconstruction operations, where reducing personnel attrition becomes an important factor.

EVACUATION

12-11. Reconditioning cases that do not recover sufficiently to return to some duty within the designated evacuation period are evacuated from corps to EAC. They are best transported in cargo trucks and buses, bus ambulances, or an ambulance train. In the latter two cases, they should be assigned helper tasks. Use air evacuation only if there is no other alternative.

SECOND-LINE RECONDITIONING

12-12. Second-line reconditioning is conducted in the intermediate staging base outside the CZ in the theater. This center could be at a CSH, a fixed MTF, or a CSC unit element. The second line reconditioning center continues to emphasize physical fitness, Soldier skills, work details, and individual/group counseling/psychotherapy. Cases will be retrained for CSS duties at corps and theater levels. As soon as the recovering Soldiers are ready, the retraining site can shift to on-the-job training at a nearby CSS unit.

THIRD-LINE RECONDITIONING

12-13. Third-line reconditioning is for Soldiers with COSR and/or mental disorders who did not improve sufficiently at lower lines of reconditioning but still have RTD potential. These Soldiers are transferred to a reconditioning program at a regional MEDCEN or home station MEDDAC. These reconditioning centers, like the others, must maintain a military atmosphere and provide opportunities to engage in occupational therapy. Some Soldiers may also require retraining for other duty/MOS reclassification.

Appendix A

Combat and Operational Stress Control and Religious Support

ROLE OF UNIT MINISTRY TEAM

A-1. This appendix addresses the general role of the UMT in the commander's program for COSC. The UMT, imbedded within units down to battalion level, provide immediate support to leaders in fulfilling their COSC responsibilities. The UMTs also assist in training leaders to recognize combat and operational stress identification and intervention responsibilities. In cooperation with unit medical personnel, UMTs serve as a primary referral agency to BH resources.

RELIGIOUS SUPPORT FOR COMBAT AND OPERATIONAL STRESS CONTROL

A-2. Soldiers' inner resources are generally rooted in their religious and spiritual values. In combat, Soldiers often show more interest in their religious beliefs. When religious and spiritual values are challenged by the chaos of combat, Soldiers may lose connection with the inner resources that have sustained them. The UMT is the primary resource available to Soldiers experiencing such dilemmas and is a valuable resource in assisting them as they seek to refocus their spiritual values.

Unit Ministry Team Support for Combat and Operational Stress Control

A-3. The UMT provides preventive, immediate, and restorative spiritual, emotional support and care to Soldiers experiencing COSR.

Preventive Religious Support for Combat and Operational Stress Reaction

A-4. The UMT assists in preventing COSR and misconduct stress disorders through spiritual fitness training. Ministry of presence with Soldiers, assigned Department of Army civilians, and contractors is critical. The UMT provides a stabilizing influence on personnel and assists them in strengthening and regaining personally held spiritual values. Preventative activities include—

- Worship opportunities.
- Private and group prayer opportunities.
- Religious literature and materials.
- Scripture readings.
- Sacraments and ordinances.
- Assistance to personnel and families prior to deployment, emphasizing family strengths.

Immediate Religious Support for Combat and Operational Stress Reaction

A-5. The UMT assists commanders in the identification of personnel experiencing negative reactions to combat and operation stress, COSR, and misconduct stress behaviors. The UMT works closely with the unit's leaders and medical personnel to care for COSR cases through religious support and comfort. Immediate religious support activities may include—

- Conversation focused upon fears, hopes, and other feelings.
- Prayer for fallen comrades and memorial ceremonies and services.
- Rites, sacraments, and ordinances, as appropriate.
- Sacred scripture.

Restorative Religious Support for Combat and Operational Stress Control

A-6. Following an operation, a unit may require reconstitution. Surviving Soldiers may need to rebuild emotional, psychological, and spiritual strength. Depending upon the spiritual, emotional, and physical condition of the unit's Soldiers, the organic UMT may need augmentation from higher echelons or other units. Restorative religious support activities may include—

- Worship, sacraments, rites, and ordinances.
- Memorial ceremonies and services.
- Religious literature and materials.
- Grief facilitation and counseling.
- Reinforcement of the Soldiers' faith and hope.
- Opportunities for Soldiers to talk about combat experiences and to integrate those experiences into their lives.

Appendix B

Medical Detachment, Combat and Operational Stress Control

CONCEPT OVERVIEW

BACKGROUND

B-1. The current medical units, CSC in the Army inventory were initially developed over ten years ago. The MF2K redesign resulted in the development of the MRI. Recent events in the Global War on Terrorism in both Afghanistan and Iraq have indicated a need to update these units from MRI medical detachment, CSC to a multifunctional COSC detachment. The 2003 US Army Surgeon General's Mental Health Advisory Team (MHAT) report and observations indicate a need to update the unit in terms of modularity, multifunctionality, and its working relationship with support units and command surgeons from brigade to theater levels.

PROPOSED MEDICAL DETACHMENT, COMBAT AND OPERATIONAL STRESS CONTROL

B-2. The proposed medical detachment, COSC (TOE 08660G000) consists of a detachment headquarters, a main support section, and a forward support section. The main support section consists of its headquarters and an 18-Soldier BH team made up of social workers, clinical psychologists, psychiatrists, occupational therapists, psychiatric nurses, MH specialists, and OT specialists. The forward support section consists of an 18-Soldier BH team only. Each BH team is capable of breaking into six 3-person subteams, for battalion/company prevention and fitness support activities. This provides for a total of 12 subteams for each detachment, giving supported commanders more teams and more flexibility in the utilization of those teams, yet maintaining all of the capabilities of COSC.

Operational Environment

B-3. Soldiers will continue to experience the physical and mental impacts of high stress in both combat and stability and reconstruction operational environments. Advances in technology which will increase effectiveness, and the impact of the individual Soldier, will also increase dispersion and add to isolation and stress levels of Soldiers. The enemy of the future will look for methods that will have the greatest psychological impact on our future Soldiers; future enemies will not be our technological equals and will increasingly rely on terror and nontraditional methods to unnerve, injure, and demoralize both Soldiers and civilians. There is an identified need to update COSC units in terms of flexibility, modularity, multifunctionality, and its relationship to supported units and the surgeons responsible to commanders for the BH mission. The current configuration of the COSC units resulted in an uneven use of available COSC resources and that there was a greater need for preventive and limited fitness services forward. Commanders have communicated there is a greater need for active outreach and consultation.

Assumptions

B-4. Power projection will likely remain the fundamental strategic and operational imperative of our forces for the foreseeable future. The new medical detachment, COSC will be well suited to the future force, designed to move with and support the division and corps of the future force, and division and corps equivalents. The medical detachment, COSC will be assigned to a MEDBDE, MMB, or MEDCOM and will be operationally attached to supported units.

B-5. Each BCT has a MH section with one BH officer and one MH specialist and that all division and brigade MH sections are filled. The medical detachments, COSC will be able to augment BCTs MH sections. The medical detachment, COSC will require additional logistical, finance, maintenance, personnel, legal, FHP, and administrative services support.

Implication for Combat and Operational Stress Control

B-6. Future forces must be emotionally and mentally fit before deploying and resilient in battle in order to endure the fast pace and intense lethality of the battlefield. Commanders will need to be able to apply COSC capabilities at every level quickly and effectively. Reorganization of the medical detachment, CSC to a combined, multifunctional medical detachment, COSC will provide greater flexibility and greatly increase outreach and preventive services while maintaining capability for the entire spectrum of COSC.

Operational Concept, Mission

B-7. The medical detachment, COSC may be attached to a theater or division medical C2 headquarters in order to provide COSC casualty prevention, treatment, and management on an area basis.

B-8. The AUTL (see FM 7-15) linkage includes—

- ART 6.5. Provide Force Health Protection in a Global Environment.
- ART 6.5.1. Provide Combat Casualty Care.
- ART 6.5.1.5. Provide Mental Health/Neuropsychiatric treatment.
- ART 6.5.4. Provide Casualty Prevention.
- ART 6.5.4.5. Provide Combat Operational Stress Control Prevention.

Method of Employment

B-9. The medical detachment, COSC deploys with a division-sized force into a theater or with a task force of up to 25,000 Soldiers in stability and reconstruction operations. The medical detachment, COSC is attached for OPCON to the division/corps/theater medical C2 headquarters. The medical detachment, COSC is assigned to the MMB of units that have reorganized under the future force design. It is normally tasked by the command surgeon to support units within the command. The detachment commander as authorized by his medical C2 headquarters will coordinate operations with the command surgeon or the division/corps/theater psychiatrist. In the absence of a psychiatrist, the detachment commander will advise and consult with the command surgeon and the commander as required. Subteams may be task-organized and further attached for operational control to maneuver BCTs, and will operate in and forward from BSA, working closely with medical assets and combat religious support team (CRST). The detachment commander will coordinate efforts with the division and the CRST to provide in-theater prevention programs as required (suicide prevention, reunion, COSC training, TEM, and PTE stress management).

Detachment Headquarters

B-10. Plans, supervises and monitors the activities of the medical detachment, COSC in support of the COSC mission. See Table B-1 for a listing of personnel assigned to the detachment headquarters. The commander will establish a command post (CP) collocated with the supported unit surgeon (either division or sustainment brigade) and coordinate all activities of the detachment through the surgeon section. The detachment commander provides C2 and works closely with supported unit surgeon to provide logistics and other required services for the detachment. The field medical assistant and/or the detachment sergeant also establishes and coordinates logistics and required services support through the supported command staff for all detachment personnel and operations.

Table B-1. Detachment headquarters personnel

Detachment Commander (LTC/05, AOC 05A00)
Medical Operations Officer (CPT/03, AOC 70B67, MS)
Chaplain (CPT/03, AOC 56A00, CH)
Detachment Noncommissioned Officer in Charge (NCOIC) (SFC/E7, MOS 68X40)
Supply Sergeant (SGT/E5, MOS 92Y20)
Human Resources Specialist (SPC/E4, MOS 42A10)
Wheeled Vehicle Mechanic (SPC/E4, MOS 63B10)
Patient Administration Specialist (SPC/E4, MOS 68G10)
Religious Support Specialist (SPC/E4, MOS 56M10)
Cook (PFC/E3, MOS 92G10)
Note. The position of detachment commander may be filled by a (LTC, AOCs 73A00, 73B00, 60W00, 65A00 or 66C00). For description of duties, see Section I of the TOE and Department of Army (DA) Pamphlet 611-21.

Main Support Section

B-11. Included in the main support section is the detachment headquarters personnel identified in Table B-1 above and an 18-personnel BH team. See Table B-2 for list of personnel assigned to the BH team. This BH team can break into 6 three-person subteams, each with one officer, one NCO, and one specialist to conduct prevention and limited fitness operations in the battalion/company level in a maneuver BSA. Teams are collocated with a BCT medical unit. Modularized subteams may be combined and task-organized to provide or support any other COSC function, which includes larger scale Soldier restoration reconstitution, or any other COSC or BH mission as determined by its headquarters. Tasks for the main support section BH team includes:

- Consultation and coordination with commanders at team level (battalion and company).
- Coordination with CRST.
- Prevention and fitness support activities.
- Reconstitute any other COSC function as required.

B-12. When attached for support to any medical company Level II MTF, the grouped subteams are dependent on the supporting medical company for logistical support and quarters. The subteam can provide 3-day restoration for up to 50 Soldiers. With additional logistical support and quarters from supported command, the team can accommodate additional Soldiers under surge conditions.

Table B-2. Main section behavioral health team

Team Leader/Psychiatrist (MAJ/04, AOC 60W00)
Occupational Therapist (MAJ/04, AOC 65A00)
Social Work Officer (MAJ/04, AOC 73A67)
Behavioral Science Officer (CPT/03, AOC 67D00)
Clinical Psychologist (CPT/03, 73B67)
Psychiatric Nurse (CPT/03, 66C7T)
Occupational Therapy NCO (Four)—(Two SSGs/E6 and Two SGTs/E5, MOS 68W30/68W20)
Mental Health NCO (Four)—(One SSG/E6 and Three SGTs/E5, MOS 68X30/68X20)
Mental Health Specialist (Six)—(SPC/E4, MOS 68X10)
Note. For description of duties, see Section I of the TOE and DA Pamphlet 611-21.

Forward Support Section

B-13. This section consists of an 18-Soldier BH team. It is comprised of a section leader and five other BH officers, six MH/OT NCOs, and six MH specialists. This BH team can break into six three-person subteams, each with one officer, one NCO, and one specialist to conduct prevention and limited fitness operations in the BCT medical company in the BSA. Modularized subteams may be combined and task-organized to provide or support any other COSC function, which includes larger scale Soldier reconstitution, or any other COSC or BH mission as determined by its headquarters. The forward support section is dependent on the supported unit for C2 and logistical support. See Table B-3 for list of personnel assigned to the forward support section BH team.

Table B-3. Forward support section behavioral health team

Team Leader/Psychiatrist (MAJ/04, AOC 60W00)
Psychiatric Nurse (MAJ/04, 66C7T)
Clinical Psychologist (MAJ/04, 73B67)
Occupational Therapist (CPT/03, AOC 65A00)
Social Work Officer (CPT/03, AOC 73A67)
Behavioral Science Officer (CPT/03, AOC 67D00)
Occupational Therapy NCO (Two)—(One SSG/E6 and One SGT/E5, MOS 68W30/68W20)
Mental Health NCO (Four)—(One SSG/E6 and Three SGTs/E5S, MOS 68X30/68X20)
Mental Health Specialist (Six)—(SPC/E4, MOS 68X10)
***Note*.** For description of duties, see Section I of the TOE and DA Pamphlet 611-21.

Organizational Concept

B-14. Required capabilities require the unit to—

- Deploy to power projection platform or point of debarkation and deploy with supported unit or deploy to theater and join supported unit.
- Provide all functions of COSC from company to theater level in close coordination with and under the direction of the supported unit surgeon.
- Provide prevention, assessment, treatment, and referral BH activities.
- Provide consultation with commanders from company to theater level.
- Conduct and coordinate administrative and logistical support to sustain operations.

Basis of Allocation Total Army Analysis

B-15. The basis of allocation is 0.333 per BCT; 1 per division; 2 per theater. In support of a theater, a COSC detachment provides support on an area basis and provides additional support to the division/corps on order. See Figure B-1 for breakdown of separate standard requirement codes (SRCs) of the medical detachment, COSC sections.

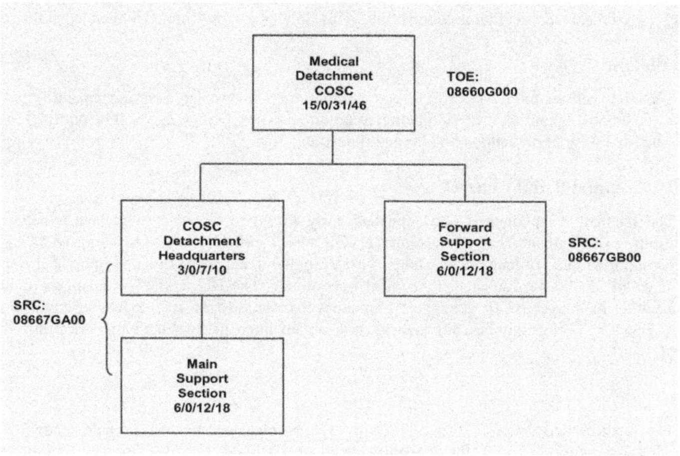

Figure B-1. Medical detachment, combat and operational stress control

Concept of Transition

B-16. Transition begins with the four active component detachments, one at a time. Personnel, transportation, and equipment changes will be minimal; transition should take a short amount of time and money to accomplish. Changes in the detachments are congruent with Force Modernization designs and personnel proponency issues. The proposed COSC detachment is designed with the Future Force in mind.

Patient Medical Records

B-17. The Surgeon General's MHAT report notes that there is a need to standardize BH reporting and documentation. The BH charts were inconsistently maintained and documentation did not always accompany patients through the evacuation chain. An additional finding was that there was no standardized method of collecting BH workload or clinical data and that no database that tracked evacuees provided reliable clinical information. The patient administration specialist (MOS 68G10) will supervise or perform functions as outlined in DA Pamphlet 611-21 for the Soldier caseload; maintain Soldier/patient accountability and records; interface with patient administration division of supported/collocated medical units.

Role of Religious Support Specialist

B-18. Findings of the MHAT include the need for chaplains to be aware of their role in COSC and that an aggressive chaplain outreach program should be executed. This requires that supervisory chaplains be involved in CRST integration with BH and primary care providers, and educate primary players in role of CRST in COSC. These activities require that the COSC chaplain be mobile and have an active presence in the maneuver brigades. This position is required to fulfill the full-time force protection requirements for the chaplain while in theater. The religious support specialist assists the COSC chaplain by screening individuals seeking counseling, coordinates for convoy security when the CRST goes out on site visits; assists chaplain with fund request; develops and coordinates needed religious support projects within the unit; coordinates with detachment for all vehicle services, communication system repairs, detachment training requirements and section accountability. This Soldier is cross-trained to develop skills for COSC.

Multifunctionality

B-19. The MH sections have a priority of prevention activities, but also have the capability to conduct or support any combat operational stress function as required. The COSC teams will be outfitted to a standard to gain the capability of prevention and limited fitness activities.

Modularity, Command and Control

B-20. The medical detachment, COSC attached early to supported units will have a better-integrated relationship. Detachment commanders and NCOICs will collocate with and coordinate all activities through supported unit surgeons. Detachments are designed to support a division in combat or a joint task force of up to 25,000 personnel in noncombat operations. The BH teams are designed to support and operate within maneuver BCTs in support of maneuver battalions for area coverage, collocating in BSA with the BSMC. Teams may be combined for task-organization to meet the entire spectrum of COSC as required.

Limitations

B-21. The medical detachment, COSC does not have its own logistics or services capability and must depend on the supported unit for logistical, legal, FHP, food service, finance, and personnel and administrative services. The COSC teams are small (one or two vehicles) and are dependent upon supported units for convoy and security operations. The logistical support necessary (space, tents, cots, and so forth) to provide Soldier restoration/reconstitution functions must be furnished by the supported element (division, corps, or theater).

Assumptions

B-22. Wartime activities will continue to cause physical and mental impacts of extremely high stress in both direct combat and the support operational environment. Advances in technology will impact the individual Soldier by increasing dispersion, and adding to their isolation and stress levels. The enemy of the future will look for methods that will have the greatest psychological impact on our future Soldier; future enemies will not be our technological equals and will increasingly rely on terror and nontraditional methods to unnerve, injure, and demoralize both Soldiers and civilians. Commanders will need to be able to quickly and effectively apply COSC capabilities at every level of the unit.

Appendix C

Medical Company and Medical Detachment, Combat Stress Control (Medical Force 2000)

SECTION I — MEDICAL COMPANY, COMBAT STRESS CONTROL (TOE 08467L000) (MEDICAL FORCE 2000)

MISSION

C-1. The mission of the AOE medical company, CSC is to provide comprehensive COSC support through directed interventions activities and COSC training for supported corps units. It provides DS to maneuver brigades lacking organic BH officers; augments units with BH assets; and provides area support. The medical company, CSC reconstitutes other COSC assets. The medical company, CSC provides COSC interventions and activities to indigenous populations as directed in stability and reconstruction operations, to include domestic support operations, humanitarian assistance, disaster relief, peace support operations, and detention facility operations. The medical company, CSC provides COSC interventions and activities to units in support of their readiness preparation and throughout their deployment cycle.

ASSIGNMENT

C-2. The AOE medical company, CSC (TOE 08467L000) is assigned to a corps MEDBDE (TOE 08422A100) or corps MEDCOM (TOE 08411A000).

EMPLOYMENT

C-3. Company headquarters usually locates with the corps medical headquarters. The employment of its teams includes their dispersal throughout the corps AO. Teams may be attached to an ASMB; a division or brigade medical company; a CSH; or other corps medical unit headquarters.

CAPABILITIES

C-4. At TOE Level 1, the medical company, CSC includes—

- Dividing the preventive section into six mobile COSC preventive teams.
- Dividing the restoration section into four mobile restoration teams with each team being equipped to hold 40 Soldiers at the same time. With additional logistical support, each team can accommodate additional Soldiers under surge conditions.
- Assisting higher headquarters BH staff with planning and coordination of COSC support, identifying the stress threat and mental and physical stressors, and the implementation of COSC functional areas.

DEPENDENCY

C-5. This unit is dependent on—

- Appropriate elements of the corps for legal; finance; field feeding; personnel and administrative services support; laundry and clothing exchange; mortuary affairs support; and security of EPW, detainee, and US prisoner patients.
- The medical headquarters to which it is assigned/attached for FHP; medical administration; logistics (including MEDLOG); medical regulating of patients; evacuation; coordination for RTD; and unit-level equipment and CE maintenance.

Mobility and Security

C-6. This unit is 100-percent mobile. It requires 100 percent of its organic personnel and equipment be transported in a single lift, using its organic vehicles. Upon relocation, Soldiers being held will require additional transportation. This unit is responsible for perimeter defense of its immediate operational area. However, it is dependent on appropriate elements of the corps for additional security, to include security of convoy operations. Personnel of the company (except the chaplain) are provided weapons for their personal defense and for the defense of their patients and/or held Soldiers.

Organization

C-7. The AOE medical company, CSC (Figure C-1) is organized into a headquarters section, a preventive section with six (modular) CSCP, and a restoration section with four (modular) combat stress control restoration (CSCR) teams. The medical company, CSC is only present in the MF2K and is an RC organization that is programmed to be in the Army inventory until the year 2010. Under the MRI a new medical detachment, CSC, TOE 08463A000, will replace this unit.

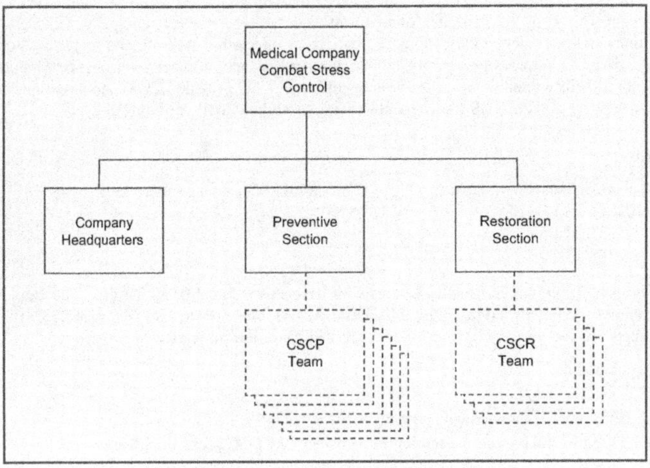

Figure C-1. Medical company, combat stress control

Company Headquarters Section

C-8. The company headquarters section provides C2 and unit-level administrative and maintenance support to its subordinate sections when they are collocated with the company. Personnel from the headquarters section are deployed with teams or task-organized COSC elements, as required. Personnel assigned to this section are identified in Table C-1.

Table C-1. Company headquarters section personnel 3/0/14

Company Commander (LTC/05, AOC 60W00, MC)
Chaplain (CPT/03, AOC 56A00, CH)
Field Medical Assistant (CPT/03, AOC 70B67, MS)
First Sergeant (1SG/E8, MOS 68W5M)
Mental Health NCO (SFC/E7, MOS 68X40)
Motor Sergeant (SSG/E6, MOS 63B30)
Supply Sergeant (SGT/E5, MOS 92Y20)
Equipment Record/Parts Sergeant (SGT/E5, MOS 92A20)
Wheeled Vehicle Mechanic (SGT/E5, MOS 63B20)
Human Resource Specialist (SPC/E4, MOS 42A10)
Nuclear, Biological, and Chemical Specialist (SPC/E4, MOS 74D10)
Wheeled Vehicle Mechanic (SPC/E4, MOS 63B10)
Unit Supply Specialist/Armorer (SPC/E4, MOS 92Y10)
Power-Generation Equipment Repairer (SPC/E4, MOS 52D10)
Cook (Three)— (Two SPC/E4 and One PFC/E3, MOS 92G10)
Note. The cook is deployed to assist the unit providing food service support to the company. He may participate in other CSC unit missions including outreach surveillance and stress control training.

Preventive Section

C-9. The 24-person preventive section staff is identified in Table C-2. The preventive section is task-organized to conduct COSC interventions and activities. Preventive section personnel may be task-organized with personnel of the COSC restoration section into teams for specific missions. The preventive section can divide into six CSCP teams. The section (and team) leader position may be held by any of the officers assigned to the section. The preventive section can augment or reconstitute the medical detachment, CSC teams.

Table C-2. Preventive section personnel 12/0/12

Psychiatrist (Three)—(MAJ/04, AOC 60W00)
Social Worker (Two)—(MAJ/04, AOC 73A67)
Psychiatrist (Three)—(CPT/03, AOC 60W00)
Social Worker (Four)—(CPT/03, AOC 73A67)
Team Chief (Six)—(SGT/E5, MOS 68X20)
Mental Health Specialist (Six)—(SPC/E4, MOS 68X10)

Restoration Section

C-10. The 44-person restoration section staff is identified in Table C-3. The restoration section may be task-organized to perform its COSC mission. Combat stress control restoration section personnel may also be task-organized with personnel of the CSCP section into teams for specific missions. This section can be divided into four CSCR teams. Each CSCR team can deploy a four-person mobile team using their HMMWV.

Table C-3. Restoration section personnel 12/0/32

Occupational Therapist (MAJ/04, AOC 65A00)
Psychiatric/Mental Health Nurse (Two)—(MAJ/04, AOC 66C00/7T)
Clinical Psychologist (MAJ/04, AOC 73B67)
Occupational Therapist (Three)—(CPT/03, AOC 65A00)
Psychiatric/Mental Health Nurse (Two)—(CPT/03, AOC 66C00/7T)
Clinical Psychologist (Three)—(CPT/03, AOC 73B67)
Mental Health NCO (Two)—(SFC/E7, MOS 68X40)
Occupational Therapy NCO (SSG/E6, MOS 68W30/N3)
Team Chief (Four)—(SSG/E6, MOS 68X30)
Patient Administration NCO (SGT/E5, MOS 68G20)
Occupational Therapy Sergeant (Two)—(SGT/E5, MOS 68W20/N3)
Mental Health NCO (Four)—(SGT/E5, MOS 68X20)
Patient Administration Specialist (Three)—(Two SPC/E4 and One PFC/E3, MOS 68G10)
Occupational Therapy Specialist (Five)—(SPC/E4, MOS 68W10/N3)
Mental Health Specialist (Ten)—(Six SPC/E4 and Four SPC/E3, MOS 68X10)

SECTION II — MEDICAL DETACHMENT, COMBAT STRESS CONTROL (TOE 08463L000) (MEDICAL FORCE 2000)

MISSION

C-11. The mission of the AOE MF2K medical detachment, CSC is to provide COSC interventions and activities to supported units in its AO. It augments division and brigade MH sections; provides direct support to combat brigades without organic BH officers; and provides area support in its AO. The medical detachment, CSC reconstitutes other brigade and division COSC assets. The medical detachment, CSC provides COSC interventions and activities to indigenous populations as directed in stability and reconstruction operations, to include domestic support operations, humanitarian assistance, disaster relief, peace support operations, and detention facility operations. The medical detachment, CSC provides COSC interventions and activities between deployments to units in support of their readiness preparation and throughout their deployment cycle. Under MRI this detachment will be replaced with medical detachment, CSC, TOE 08463A000.

> ***Note.*** The MF2K and MRI medical detachments, CSC conduct COSC support operations very similarly. However, in the MRI medical detachment, CSC, psychiatry assets are assigned to the restoration section and psychology assets are assigned to the preventive section whereas in the MF2K detachment, the manning strategy for psychology and psychiatry is reversed: psychologists are in the restoration section and psychiatrists are on the preventive teams.

ASSIGNMENT

C-12. The MF2K medical detachment, CSC (TOE 08463L000), is assigned to a corps MEDBDE (TOE 08422A100), or a corps MEDCOM (TOE 08411A000), or other medical task force C2 elements. Its teams may be attached to a medical company, CSC (TOE 08467L000); an ASMB (TOE 08456A000); a brigade or divisional medical company; or a CSH.

EMPLOYMENT

C-13. Detachment headquarters is usually located in the DSA. Its teams disperse and are employed throughout its AO.

CAPABILITIES

C-14. The MF2K medical detachment, CSC provides—

- Mobile COSC interventions from the preventive section which can divide into three preventive teams.
- Holding for 40 Soldiers by the restoration team for COSC interventions such as restoration. With additional logistical support, the team can accommodate additional Soldiers under surge conditions.

STAFF RESPONSIBILITIES

C-15. The medical detachment assists the C2 headquarters (to which it is assigned or attached) regarding planning and coordination of COSC support, stress threat, mental and physical stressors, stress behaviors, principles of COSC, and implementation of COSC functional areas.

DEPENDENCY

C-16. This unit is dependent on—

- Appropriate elements of the supporting unit for FHP; religious support; legal; finance; field feeding; personnel and administrative services support; laundry and clothing exchange; mortuary affairs support; and security of EPW, detainee, and US prisoner patients.
- Supporting unit for medical administration; logistics (including MEDLOG); medical regulating; evacuation; coordination for RTD; and unit-level equipment and CE maintenance.

MOBILITY AND SECURITY

C-17. This unit is 100-percent mobile. It requires 100 percent of its organic personnel and equipment be transported in a single lift, using its organic vehicles. Upon relocation, Soldiers being held will require additional transportation. This unit is responsible for perimeter defense of its immediate operational area. However, it is dependent on appropriate elements of the supporting unit for additional security, to include security of convoy operations. Personnel of the detachment are provided weapons for their personal defense and for the defense of their patients and/or held Soldiers.

ORGANIZATION

C-18. This 25-person unit (see Table C-4) is composed of a headquarters, a combat stress preventive section with three preventive teams, and a CSCR. The modular CSC teams found in the MF2K CSC medical detachment are similar to those found in the CSC medical company.

Table C-4. Detachment teams

Headquarters	*Preventive Section*	*Restoration Section*
	CSCP#1	CSCR
	CSCP#2	
	CSCP#3	

C-19. The detachment headquarters provides C2 for the detachment. It is responsible for planning, coordinating, and implementing COSC interventions and activities for supported units. It consists of three personnel: a detachment Commander (05, 60W00, MC), a detachment sergeant (E7, 68X40), and a wheeled-vehicle mechanic (E4, 63B10). The detachment commander also serves as a treating physician with the preventive section. The detachment NCOIC (a senior MH NCO) also serves as the restoration team sergeant. Detachment officers and NCOs from the prevention team and the restoration team may be assigned additional duties, which enhance the overall effectiveness of the headquarters section. See Table C-5 for personnel assigned to detachment headquarters.

Table C-5 Detachment headquarters personnel

Commander (LTC/05, AOC 60W00)
Detachment Sergeant (SFC/E7, MOS 68X40)
Wheeled Vehicle Mechanic (SPC/E4, MOS 63B10)
Note. The commander also serves as the psychiatrist in the preventive section.

Preventive Section

C-20. The 12-person preventive section staff is identified in Table C-6. The preventive section may be task-organized to conduct its COSC mission. Preventive section personnel may also be task-organized with personnel of the restoration section into teams for specific missions. The preventive section can divide into three preventive teams. The section (and team) leader position may be held by any of the officers assigned to the section. This section's COSC interventions and activities are—

- Unit needs assessment; consultation and education; critical event and transition management; COSC triage; stabilization (emergency); and BH treatment.
- Assisting with restoration and reconditioning at the CSC detachment program. Overseeing a 1- to 3-day COSC restoration program in a brigade, division, or ASMC holding section or in another area suitable for Soldiers experiencing COSR and/or other stress-related disorders.

Table C-6. Preventive section 7/0/6

Psychiatrist (LTC/05, AOC 60W00)
Clinical Psychologist (Three)—(MAJ/04, AOC 73B67)
Social Worker (Three)—(CPT/03, AOC 73A67)
Team Chief (Three)—(SGT/E5, MOS 68X20)
Mental Health Specialist (Three)—(SPC/E4, MOS 68X10)
Note. The psychiatrist is also counted in the headquarters section.

Restoration Team

C-21. The 9-person restoration team staff is identified in Table C-7. The restoration team is task-organized to provide COSC interventions and activities. Restoration team personnel may also be task-organized with personnel of the CSC preventive section into teams for specific missions. Each restoration team can deploy a four-person mobile team with a HMMWV. The CSCF usually collocates with a supported divisional medical company to provide mobile COSC support within a DSA and conduct restoration programs, as required. The CSCF provides staff and equipment for operating a restoration or reconditioning center. This section's COSC interventions and activities are—

- Unit needs assessment; consultation and education; critical event and transition management; COSC triage; stabilization; and BH treatment.
- Conducting Soldier restoration and reconditioning programs.
- Assisting the CSH psychiatric section when psychiatric ward capability is required (refer to FM 8-10-14).

Table C-7. Restoration team 3/0/6

Psychiatrist (MAJ/04, AOC 60W00)
Psychiatric/Mental Health Nurse (MAJ/04, AOC 66COO/7T)
Occupational Therapist (CPT/03, AOC 65A00)
Occupational Therapy NCO (SSG/E6, 68W30/N3)
Team Chief (SSG/E6, MOS 68X30)
Occupational Therapy Sergeant (SGT/E5, MOS 68W20/N3)
Mental Health NCO (SGT/E5, MOS 68X20)
Mental Health Specialist (Two)—(One SPC/E4 and One PFC/E3, MOS 68X10)

Glossary

SECTION I — ACRONYMS AND ABBREVIATIONS

AAR	after-action review
ABCA	American, British, Canadian, and Australian
ACR	armored cavalry regiment
AMEDD	Army Medical Department
AN	Army Nurse Corps
AO	area of operations
AOC	area of concentration
AOE	Army of Excellence
AR	Army regulation
ART	Army tactical task
ASMB	area support medical battalion
ASMC	area support medical company
AUTL	Army Universal Task List
BAS	battalion aid station
BCT	brigade combat team
BDP	behavioral disordered patient
BF	battle fatigue
BH	behavioral health
BICEPS	brevity, immediacy, contact, expectancy, proximity, and simplicity (See Section II for definition.)
BSA	brigade support area
BSB	brigade support battalion
BSMC	brigade support medical company
C	Centigrade
C2	command and control
C4I	command, control, communications, computers, and intelligence
CBRN	chemical, biological, radiological, and nuclear
CE	communications-electronics
CH	chaplain
CHL	combat health logistics
CHS	combat health support
COA	course(s) of action
CONUS	continental United States
COSC	combat and operational stress control (See Section II for definition.)
COSR	combat and operational stress reaction (See Section II for definition.)

CP	command post
CPT	captain
CRST	combat religious support team
CS	combat support
CSC	combat stress control
CSCF	combat stress control fitness
CSCP	combat stress control preventive
CSCR	combat stress control restoration
CSH	combat support hospital
CSR	combat stress reaction
CZ	combat zone
DA	Department of the Army
DD	Department of Defense
DEPMEDS	Deployable Medical Systems
DISCOM	division support command
DMHS	division mental health section
DNBI	disease and nonbattle injury(ies)
DOD	Department of Defense
DODD	Department of Defense Directive
DODI	Department of Defense Instruction
DS	direct support
DSA	division support area
EAB	echelons above brigade
EAC	echelons above corps
EAD	echelons above division
EMT	emergency medical treatment
EPW	enemy prisoner(s) of war
ETOD	end-of-tour debriefing
F	Fahrenheit
FAP	Family Advocacy Program
FHP	force health protection
1LT	first lieutenant
1SG	first sergeant
FM	field manual
FMC	United States Field Medical Card (Department of Defense Form 1380)
FRAGO	fragmentary order
FSB	forward support battalion
FSMC	forward support medical company
G-1	Assistant Chief of Staff (Personnel)
G-2	Assistant Chief of Staff (Intelligence)
G-3	Assistant Chief of Staff (Operations and Plans)

G-4	Assistant Chief of Staff (Logistics)
G-5	Assistant Chief of Staff (Plans and Policy)
G-9	Assistant Chief of Staff (Civil-Military Affairs)
GP	general purpose
HBCT	heavy brigade combat team
HHD	headquarters and headquarters detachment
HIP	help-in-place
HMMWV	high-mobility multipurpose wheeled vehicle
HSL	health service logistics
HSS	health service support
IED	improvised explosive device
JAG	Judge Advocate General
LTC	lieutenant colonel
MAJ	major
MC	Medical Corps
MCW	minimal care ward
MEDBDE	medical brigade
MEDCEN	medical center
MEDCOM	medical command (US Army)
MEDDAC	medical department activity
MEDLOG	medical logistics
METT-TC	mission, enemy, terrain and weather, troops and support available, time available, and civil considerations
MF2K	Medical Force 2000
mg	milligram(s)
MH	mental health
MHAT	Mental Health Advisory Team (US Army)
MMB	multifunctional medical battalion
MOS	military occupational specialty
MP	military police
MRI	Medical Reengineering Initiative
MRO	medical regulating officer
MS	Medical Service Corps
MSB	main support battalion
MSMC	main support medical company
MTF	medical treatment facility
MTOE	modification table of organization and equipment
NATO	North Atlantic Treaty Organization
NCO	noncommissioned officer
NCOIC	noncommissioned officer in charge
NP	neuropsychiatric

OEF	Operation Enduring Freedom
OIF	Operation Iraqi Freedom
OPCON	operational control
OPLAN	operation plan
OPORD	operation order
OPTEMPO	operational tempo
OSAT	operational stress assessment team
OSR	operational stress reaction
OT	occupational therapy
PA	physician assistant
PCP	phencyclidine hydrochloride
PD	psychological debriefing
PERSTEMPO	personnel tempo
PIES	proximity, immediacy, expectancy, and simplicity
POC	point of contact
PTE	potentially traumatizing event
PTSD	posttraumatic stress disorder
pub	publication
PVNTMED	preventive medicine
R&R	rest and relaxation
RC	Reserve Component
RTD	return to duty
S1	Adjutant (US Army)
S2	Intelligence Officer (US Army)
S3	Operations and Training Officer (US Army)
S4	Supply Officer (US Army)
SBCT	Stryker Brigade Combat Team
SGT	sergeant
SOI	signal operation instructions
SOP	standing operating procedure
SP	Army Medical Specialty Corps
SPC	specialist
SRC	standard requirement code
SSG	staff sergeant
TC	training circular
TDA	table of distribution and allowances
TEM	traumatic events management
TEMPER	tent, expandable, modular, personnel
TG	technical guide
TOE	table(s) of organization and equipment
TRADOC	United States Army Training and Doctrine Command

TSOP	tactical standing operating procedures
UCMJ	Uniform Code of Military Justice
UMT	unit ministry team
UNA	unit needs assessment
US	United States
USAMEDDC&S	United States Army Medical Department Center and School
USAMRMC	United States Army Medical Research and Materiel Command
USAR	United States Army Reserve
USMC	United States Marine Corps
VHA	Veterans Health Adminstration

SECTION II — TERMS

brevity, immediacy, contact, expectancy, proximity, and simplicity

An acronym used for the management of combat and operational stress reactions—*brevity* (usually less than 72 hours); *immediacy* (as soon as symptoms are evident); *contact* (chain of command remains directly involved in the Soldiers recovery and return to duty), *expectancy* (combat stress control unit personnel expectation that casualties will recover); *proximity* (of treatment at or as near the front as possible); *simplicity* (the use of simple measures such as rest, food, hygiene, and reassurance). Also known as BICEPS.

combat and operational stress control

A coordinated program for the prevention of and actions taken by military leadership to prevent, identify, and manage adverse combat and operational stress reactions in units. Also known as COSC.

combat and operational stress reaction

The expected, predictable, emotional, intellectual, physical, and/or behavioral reactions of Service members who have been exposed to stressful events in combat or military operations other than war. Also known as COSR.

5 R's

Actions used for combat and operational stress reaction control that include—**R**eassure of normality; **R**est (respite from combat or break from the work); **R**eplenish bodily needs (such as thermal comfort, water, food, hygiene, and sleep); **R**estore confidence with purposeful activities and contact with his unit; **R**eturn to duty and reunite Soldier with his unit.

reconditioning program

An intensive 4- to 7-day program (may be extended by exception to theater evacuation policy) of replenishment, physical activity, therapy, and military retraining for combat and operational stress control casualties and neuropsychiatric cases (including alcohol and drug abuse) who require successful completion for return to duty or is evacuated for further neuropsychiatric evaluation.

Soldier restoration

A 24- to 72-hour (1- to 3-day) program in which Soldiers with combat and operational stress reactions receive treatment.

stabilization

The initial short-term management and evaluation of severely behaviorally disturbed Soldiers caused by an underlying combat and operational stress reaction, behavioral health disorder, or alcohol and/or drug abuse reaction.

References

SOURCES USED

These are the sources quoted or paraphrased in this publication.

DEPARTMENT OF DEFENSE

These documents are available online at: http://www.dtic.mil/whs/directives/

DODD 6490.1, *Mental Health Evaluations of Members of the Armed Forces*, 1 October 1997

DODD 6490.2, *Comprehensive Health Surveillance*, 21 October 2004

DODD 6490.5, *Combat Stress Control (CSC) Program*, 23 February 1999

DODI 6490.4, *Requirements for Mental Health Evaluations of Members of the Armed Forces*, 28 August 1997

MULTISERVICE PUBLICATION

FM 6-22.5/MCRP 6-11C/NTTP 1-15M, *Combat Stress*, 23 June 2000

ARMY PUBLICATIONS

These publications are available online at: https://akocomm.us.army.mil/usapa

AR 40-216, *Neuropsychiatry and Mental Health,* 10 August 1984

FM 7-15, *The Army Universal Task List*, 31 August 2003

FM 22-51, *Leaders' Manual for Combat Stress Control,* 29 September 1994

DOCUMENTS NEEDED

These documents must be available to the intended users of this publication.

DEPARTMENT OF DEFENSE FORMS

These forms are available online at: https://akocomm.us.army.mil/usapa

DD Form 1380, *US Field Medical Card*

DD Form 2795, *Pre-Deployment Health Assessment*

DD Form 2796, *Post-Deployment Health Assessment*

MULTISERVICE PUBLICATION

AR 190-8/OPNAVINST 3461.6/AFJI 31-304/MCO 3461.1, *Enemy Prisoners of War, Retained Personnel, Civilian Internees, and Other Detainees,* 1 October 1997

ARMY PUBLICATIONS

These publications are available online at: https://akocomm.us.army.mil/usapa

AR 25-1, *Army Knowledge Management and Information Technology Management*, 15 July 2005

AR 40-3, *Medical, Dental, and Veterinary Care,* 12 November 2002

AR 40-61, *Medical Logistics Policies*, 28 January 2005

AR 40-66, *Medical Record Administration and Health Care Documentation,* 20 July 2004

AR 40-68, *Clinical Quality Management,* 26 February 2004

AR 40-400, *Patient Administration*, 12 March 2001

AR 40-501, *Standards of Medical Fitness,* 1 February 2005

AR 340-21, *The Army Privacy Program*, 5 July 1985

DA Pamphlet 611-21, *Military Occupational Classification and Structure*, 31 March 1999

FM 4-02, *Force Health Protection in a Global Environment,* 13 February 2003

FM 4-02.6, *The Medical Company—Tactics, Techniques, and Procedures,* 1 August 2002 (Change 1, 9 April 2004)

FM 4-02.7, *Health Service Support in a Nuclear, Biological, and Chemical Environment—Tactics, Techniques, and Procedures,* 1 October 2002

FM 4-02.10, *Theater Hospitalization*, 3 January 2005

FM 4-02.16, *Army Medical Information Management—Tactics, Techniques, and Procedures,* 22 August 2003

FM 4-02.17, *Preventive Medicine Services,* 28 August 2000

FM 4-25.12, *Unit Field Sanitation Team,* 25 January 2002

FM 8-10-6, *Medical Evacuation in a Theater of Operations—Tactics, Techniques, and Procedures,* 14 April 2000

READINGS RECOMMENDED

These sources contain relevant supplemental information.

JOINT OR MULTISERVICE PUBLICATIONS

Most joint publications are available online at: http://www.dtic.mil/doctrine/

Joint Pub (JP) 1-02, *Department of Defense Dictionary of Military and Associated Terms,* 12 April 2001 (As amended through 9 June 2004)

JP 3-0, *Doctrine for Joint Operations*, 9 September 2001

JP 4-02, *Doctrine for Health Service Support in Joint Operations*, 30 July 2001

FM 1-02/MCRP 5-12A, *Operational Terms and Graphic*, 21 September 2004

FM 4-25.11/NTRP 4-02.1.1/AFMAN 44-163(I)/MCRP 3-02G), *First Aid,* 23 December 2002 (Change 1, 15 July 2004)

FM 8-9/NAVMED P-5059/AFJMAN 44-151V1V2V3, *NATO Handbook on the Medical Aspects of NBC Defensive Operations AMedP-6(B), Part I—Nuclear, Part II—Biological, Part IIII—Chemical,* 1 February 1996

FM 8-284/NTRP 4-02.23(NAVMED P-5042)/AFMAN (I) 44-156/MCRP 4-11.1C, *Treatment of Biological Warfare Agent Casualties,* 17 July 2000 (Change 1, 8 July 2002)

FM 8-285(4-02.285)/NAVMED P-5041/AFJMAN 44-149/FMFM 11-11, *Treatment of Chemical Agent Casualties and Conventional Military Chemical Injuries,* 22 December 1995

FM 21-10/MCRP 4-11.1D, *Field Hygiene and Sanitation,* 21 June 2000

Army Publications

Most of these publications are available online at: https://akocomm.us.army.mil/usapa

AR 71-32, *Force Development and Documentation—Consolidated Policies*, 3 March 1997

AR 635-200, *Active Duty Enlisted Administrative Separations*, 6 June 2005

FM 3-0, *Operations*, 14 June 2001

FM 3-06.11, *Combined Arms Operations in Urban Terrain*, 28 February 2002

FM 3-07, *Stability Operations and Support Operations*, 20 February 2003

FM 3-19.40, *Military Police Internment/Resettlement Operations*, 1 August 2001

FM 3-21.21, *The Stryker Brigade Combat Team Infantry Battalion*, 8 April 2003 (Change 1, 31 July 2003)

FM 3-21.31, *The Stryker Brigade Combat Team*, 13 March 2003

FM 3-90.3, *The Mounted Brigade Combat Team*, 1 November 2001

FM 5-0, *Army Planning and Orders Production*, 20 January 2005

FM 4-0, *Combat Service Support*, 29 August 2003

FM 4-02.4, *Medical Platoon Leaders' Handbook—Tactics, Techniques, and Procedures*, 24 August 2001 (Change 1, 18 December 2003)

FM 4-02.19, *Dental Service Support in the Theater of Operations*, 1 March 2001

FM 4-02.21, *Division and Brigade Surgeons' Handbook (Digitized)—Tactics, Techniques, and Procedures*, 15 November 2000

FM 4-02.24, *Area Support Medical Battalion— Tactics, Techniques, and Procedures*, 28 August 2000

FM 4.25.10, *Field Hygiene and Sanitation*, 20 June 2002

FM 8-10-14, *Employment of the Combat Support Hospital Tactics, Techniques, and Procedures*, 29 December 1994 (will be revised as FM 4-02.14)

FM 8-42, *Combat Health Support in Stability Operations and Support Operations*, 27 October 1997 (will be revised as FM 4-02.42)

FM 8-50, *Prevention and Medical Management of Laser Injuries*, 8 August 1990 (will be revised as FM 4-02.50)

FM 8-55, *Planning for Health Service Support,* 9 September 1994 (will be revised as FM 4-02.55)

FM 10-27-4, *Organizational Supply and Services for Unit Leaders,* 14 April 2000 (will be revised as FM 4-20.05)

FM 27-1, *Legal Guide for Commanders.* 13 January 1992 (will be revised as FM 1-04.1)

FM 27-10, *The Law of Land Warfare,* 18 July 1956 (Reprinted with basic including Change 1, 15 July 1976) (will be revised as FM 1-04.10)

FM 63-20, *Forward Support Battalion*, 26 February 1990 (will be revised as FM 4-93.20)

FM 63-21, *Main Support Battalion*, 7 August 1990 (will be revised as FM 4-93.21)

FM 100-9, *Reconstitution*, 13 January 1992 (will be revised as FM 4-100.9)

FM 100-17, *Mobilization, Deployment, Redeployment, and Demobilization,* 28 October 1992 (will be revised as FM 3-35)

FM 100-17-5, *Redeployment*, 29 September 1999 (will be revised as FM 3-35)

MEDCOM Regulation 40-38, *Command-Directed Mental Health Evaluation*, 1 September 2001

TC 25-20, *A Leader's Guide to After-Action Reviews,* 30 September 1993

TRADOC Pamphlet 600-22, *Leaders Guide for Suicide Prevention Planning,* 16 February 2005 (Available at http://www.tradoc.army.mil/publications.htm)

United States Army Center for Health Promotion and Preventive Medicine

Most of these publications are available online at: http://chppm-www.apgea.army.mil/dhpw/

USACHPPM TG 240, *Combat Stress Behaviors,* June 2004

USACHPPM TG 241, *Combat Operational Stress Reaction* (*COSR*) (*"Battle Fatigue"*), June 2004

USACHPPM TG 242, *Combat Operational Stress Reactions* (*COSR*) *Prevention: Leader Actions,* June 2004

USACHPPM TG 243, *Combat Stress Card,* May 1999

Tables of Organization and Equipment

TOE 08411A000, *Medical Command, Corps* (*MRI*)

TOE 8422A100, *Medical Brigade, Corps* (*MRI*)

TOE 08456A000, *Area Support Medical Battalion*

TOE 08457A000, *Area Support Medical Company*

TOE 08463A000, *Medical Detachment, Combat Stress Control* (*MRI*)

TOE 08463L000, *Medical Detachment, Combat Stress Control* (*MF2K*)

TOE 08467L000, *Medical Company, Combat Stress Control* (*MF2K*)

TOE 08660G000, *Medical Detachment, Combat Operational Stress Control* (*Force Development Update*)

DEPARTMENT OF VETERANS AFFAIRS

Veterans Health Administration, *Clinical Practice Guidelines (Mental Health),* available online at: http://www.oqp.med.va.gov/cpg

Index

References are to paragraph numbers except where otherwise specified.

Made in the USA
Monee, IL
05 November 2026

67810977R00164